AF781057

PHYSICS RESEARCH AND TECHNOLOGY

ADVANCED HIGH TC FERROELECTRICS

PHYSICS RESEARCH AND TECHNOLOGY

Additional books in this series can be found on Nova's website under the Series tab.

Additional e-books in this series can be found on Nova's website under the e-book tab.

PHYSICS RESEARCH AND TECHNOLOGY

ADVANCED HIGH TC FERROELECTRICS

JINGZHONG XIAO

nova publishers
New York

Copyright © 2013 by Nova Science Publishers, Inc.

All rights reserved. No part of this book may be reproduced, stored in a retrieval system or transmitted in any form or by any means: electronic, electrostatic, magnetic, tape, mechanical photocopying, recording or otherwise without the written permission of the Publisher.

For permission to use material from this book please contact us:
Telephone 631-231-7269; Fax 631-231-8175
Web Site: http://www.novapublishers.com

NOTICE TO THE READER

The Publisher has taken reasonable care in the preparation of this book, but makes no expressed or implied warranty of any kind and assumes no responsibility for any errors or omissions. No liability is assumed for incidental or consequential damages in connection with or arising out of information contained in this book. The Publisher shall not be liable for any special, consequential, or exemplary damages resulting, in whole or in part, from the readers' use of, or reliance upon, this material. Any parts of this book based on government reports are so indicated and copyright is claimed for those parts to the extent applicable to compilations of such works.

Independent verification should be sought for any data, advice or recommendations contained in this book. In addition, no responsibility is assumed by the publisher for any injury and/or damage to persons or property arising from any methods, products, instructions, ideas or otherwise contained in this publication.

This publication is designed to provide accurate and authoritative information with regard to the subject matter covered herein. It is sold with the clear understanding that the Publisher is not engaged in rendering legal or any other professional services. If legal or any other expert assistance is required, the services of a competent person should be sought. FROM A DECLARATION OF PARTICIPANTS JOINTLY ADOPTED BY A COMMITTEE OF THE AMERICAN BAR ASSOCIATION AND A COMMITTEE OF PUBLISHERS.

Additional color graphics may be available in the e-book version of this book.

Library of Congress Cataloging-in-Publication Data

Xiao, Jingzhong.
Advanced high Tc ferroelectrics / Jingzhong Xiao (Physics Department, Universidade de Coimbra, Coimbra, Portugal).
pages cm
Includes bibliographical references and index.
ISBN: 978-1-62618-002-4 (hardcover)
1. Feroelectricity. 2. Ferroelectric devices. 3. Ferroelectric crystals. I. Title.
QC596.5.X53 2013
548'.85--dc23

2013004713

Published by Nova Science Publishers, Inc. † New York

To the immortal memory of my first son, Boas Xiao

CONTENTS

PREFACE

Ferroelectrics are technologically important materials, particularly for their applications in ferroelectric random access memory (FeRAM) and Micro-Electro-Mechanical Systems (MEMS). MEMS) is the integration of mechanical elements, sensors, actuators, and electronics on a common silicon substrate through microfabrication technology. Electromechanical actuators directly transform input electrical energy into mechanical energy. Piezoelectric and electrostrictive ceramics are widely used in applications requiring high generative force, high frequency operation, accurate displacement, quick response time, or small device size. Recent developments in regard to relaxor-based single crystal piezoelectrics such as $Pb(Zn_{1/3}Nb_{2/3})O_3$– $PbTiO_3$ and $Pb(Mg_{1/3}Nb_{2/3})O_3$– $PbTiO_3$ have shown superior performance characteristics compared to $Pb(Zr,Ti)O_3$ (PZT) polycrystalline ceramics. However, their low Curie temperature (Tc; 170 °C) and relatively low ferroelectric phase transition temperatures (rhombohedral to tetragonal phase) (Tr, ~100 °C) limit their implementation in many fields. In most applications, especially in the automotive and aerospace industries, the need for actuation and sensing over a broad temperature range is essential. Relaxor ferroelectrics, such as $BiScO_3$-$PbTiO_3$ (BSPT) systems, have high Curie temperatures and hold excellent properties which meet the requirements in applications as next generation materials for actuators in automotives and aerospace sensors. In comparison with perovskite ferroelectrics such as lead zirconate titanate (PZT), this bismuth based ferroelectrics offer several advantages such as fatigue free, low operating voltages, large remanent polarization and high Curie temperature.

This book presents the important development for the studies in the ferroelectrics with a high Curie temperature (Tc), consisting six chapters. This book deals with the preparation, structural details and electrical behavior of a few peroskites phase bismuth based ferroelectrics.

Chapter 1 consists of the contents as detailed as a general introduction about ferroelectric materials and their properties. Ferroelectrics are materials that display an electric polarization in the absence of an external applied electric field, together with the property that the direction of the polarization may be reversed by an electric field. The general property knowledge of ferroelectrics, such as polarization, spontaneous polarization, ferroelectric domain, ferroelectric hysteresis and ferroelectric phase transition are introduced.

Chapter 2 provides a simple introduction for relaxor ferroelectrics and the fundamentals of high Curie temperature ferroelectrics. In the part of introducing relaxors, the knowledge and theories for treating the relaxor issues are addressed; especially, the advanced analysis methods based on synchrotron radiation sources are presented. This chapter also gives a simple introduction for the high Curie temperature ferroelectrics.

Chapter 3 focuses on the high Tc ferroelectric ceramics. There appeared a large quantity of achievements in development of high Tc ceramics in recent years, this chapter mainly summarized the researches of Bi-based piezoelectric ceramics, which are very important because of their excellent properties.

Chapter 4 indicated clearly the development for the High-Tc ferroelectric crystals, particularly Bi based crystals. The properties of several high Curie relaxor based ferroelectric single crystals such as PYNT, BSPT and BS-BG-PT with the high Cutie temperature were summaried. To achieve high performance with high levels of piezoelectric activity over a wide temperature range, the dopant modifications and other ways were used to enhance the Curie temperature at a certain extent and to develop more effective materials and satisfy the needs of modern industry.

Chapter 5 provides a simple introduction to nanoscience and nanotechnology for ferroelectric materials. This chapter indicates the development of nano ferroelctrics with a high Curie temperature, and summarize the ferroelectric properties in nano scale.

Chapter 6 introduces the research details that aimed at the development of a reliable processing method for preparation of alternative relaxor ferroelectrics with high Curie temperature and high orientation (textured). This research activity had been mainly performed by the author and the collaborations, and focused on the chemical solution fabrication and characterization of $BiScO_3$-$PbTiO_3$ (BS-PT) thin film system.

During the preparation of this book, the PhD student of the author, Miss **Lan Xu**, contributed greatly to the writing of Chapter 3; all the other chapters were written by the author (Dr. Jingzhong Xiao).

This academic book publication was supported from funds from FEDER (Programa Operacional Factores de Competitividade COMPETE), FCT - Fundação para a Ciência e Tecnologia underthe project PEst-C/FIC/UI0036/2011, FCT R&D project of PTDC/FIS/116146/2009, FCT Science/2008 program, and FCT post-doc grant of SFRH/BPD/20265/2004.

Dr. Jingzhong Xiao

CEMDRX,
Physics Department,
Rua Larga
University of Coimbra
P-3004 516 Coimbra
Portugal

Chapter 1

INTRODUCTION: FUNDAMENTALS FOR FERROELECTRICS

1.1. AN EARLY DEVELOPMENT HISTORY OF FERROELECTRIC MATERIALS

The physical properties of Rochelle salt began to attract interest in the nineteenth century. The phenomenon of pyroelectricity in various crystals had been observed by David Brewster in 1824. The existence of the piezoelectric effect in Rochelle Sal was identified by the brothers Pierre and Paul-Jacques Curie in 1880 [1]. As piezoelectrics, Rochelle Salt and other crystals were far more active than quartz, for instance. At the time of Rochelle salt, piezoelectrics were of purely academic significance.

In 1914-18, during World War I, [2, 3] A. Nicholson and Paul Langevin independently invented similar ultrasonic submarine detectors, by a mosaic of thin quartz crystals glued between two steel plates with a resonant frequency of about 50 KHz. These transducers had been mounted in a housing suitable for submersion. At the end of this war, the transducers they invented had the ability of emitting a high frequency underwater wave and measuring the depth by timing the return echo. Since that time, the development of sonar transducers, circuits, systems, and materials has continued, and the physicists and electrical engineers showed an increasing interest in the physical properties mainly because of the unusually high piezoelectric moduli in the quartz and other piezoelectric materials, due to the strategic importance of these achievements.

A professor of Theoretical Physics at the University of Zürich, Petrus Josephus Wilhelmus Debye (Peter Debye as we know him today), had carefully studied the piezoelectricity phenomena, and proposed an idea in 1912, to explain the results that a certain class of molecules carry a permanent electric dipole moment in analogy to the magnetic moment of the atoms of paramagnetic substances. The following is the equation Debye gave for Langevin's theory of paramagnetism:

$$(\varepsilon-1)/(\varepsilon+2)=a+b/T,$$

where a is proportional to the density of the substance and b to the square of the electric dipole moment [4], which was perfectly confirmed in many other systems.

According to this relationship for a critical temperature *Tc=b/(1-a)*, Debye drew a further conclusion that the dielectric constant reaches infinity. This critical temperature was considered to be an analogue to the Curie temperature of a ferromagnet. Below *T*c, a permanent dielectric polarization could be expected even in the absence of an electric field. The basic feature of ferroelectricity was anticipated based on such a phenomenon.

In fact, the term ferroelectric or ferroelectricity was promoted by Schrödinger as early as 1912. In this year, E. Schrödinger went a further step that elaborated on Debye's simple model and tried to extend it to solids, in his "Habilitations-Schrift" submitted at the University of Vienna, a year after Debye published his "Vorläufige Mitteilung". He speculated that all solids should become "ferroelektrisch" at a sufficiently low temperature, if this could be done successfully. [5]

During the early Rochelle Salt period, Joseph Valasek was another major contributor to the development of ferroelectricity sciences. He was born on April 27, 1897 in Cleveland, Ohio and received his bachelor's degree in physics at Case Institute of Technology. He firstly worked at the National Bureau of Standards in Washington D.C., where he became familiar with various methods of pyrometry applied in the steel industry and studied the ferromagnetism phenomenon both theoretically and experimentally. In 1919, he got a graduate teaching assistant position at the University of Minnesota in Minneapolis. Under the supervision of Prof. W. F. G. Swann, a famous physicist in the fields of theoretical work in cosmic rays physics and also known to have a wide knowledge in many other areas, J. Valasek began to systematically study the analogy between the magnetic properties of ferromagnetics and the dielectric properties of Rochelle Salt.

In April 1920, J. Valasek became a Master of Arts, and made a presentation with his first report on "Piezoelectric and allied Phenomena in Rochelle salt", at the meeting of the American Physical Society in Washington, in which he proposed that the dielectric displacement (D), electric intensity (E), and polarization (P) are analogous to B, H and I in the case of magnetism, and the hysteresis in P was analogous to magnetic hysteresis. This would suggest a parallelism between the behavior of Rochelle salt as a dielectric and ferromagnetic substance. With this idea, the typical hysteresis curves were obtained for Rochelle salt, analogous to the B, H curves of magnetism. His presentation was submitted to Physical Review in December 1920 [6], which marked two milestones. Firstly, J. Valasek stated for the first time that permanent polarization is the natural state of Rochelle salt, and secondly he published the first hysteresis curve of a ferroelectric material.

In 1922, Valasek completed his Ph.D. Thesis with the title of "Piezo-electric activity of Rochelle salt under various conditions" [7]. The most valuable graph in this thesis is the curve of the temperature dependence of piezoelectric response, indicating the existence of a relatively narrow temperature range of high piezoelectric activity and the existence of two phase transitions. In fact, it was Valasek, that mentioned for the first time the existence of two Curie points in Rochelle salt, although he never used the word ferroelectricity (maybe he never heard about Erwin Schrödingers proposal). But, his studies later led to the firm establishment of the term "ferroelectricity" to describe this set of phenomena.

After Rochelle salt, scientists continued to search whether there are more other materials with ferroelectric characteristics. Among them Paul Scherrer, together with Peter Debye, worked at the "Physikalisches Institut" at the "Eidgenössische Technische Hochschule" (ETH) in Zürich on dielectrics. They believed that the molecules of water of crystallization were essential for a ferroelectric material. He and his student Georg Busch searched the literature by Gert Steulmann at the "Institut für allgemeine Elektrotechnik" of the "Technische Hochschule" in Dresden [8], in which they found that various potassium salts exhibited average dielectric constants that had been measured by an immersion method, among others K_3PO_4, K_2HPO_4 and KH_2PO_4.

For K_3PO_4 and K_2HPO_4, they observed normal values of 7.75 and 9.05 respectively, but for KH_2PO_4 a value was found to be higher than 30. In view of the orthorhombic symmetry of potassium-dihydrogen phosphate, this supported Busch's vague idea that mobile hydrogen atoms might be responsible for ferroelectricity, rather than the water molecules – the hypothesis of his supervisor Paul Scherrer. Indeed, J. West at Bragg's laboratory in Manchester found that the O-H-O bonds do exist in KH_2PO_4 by a quantitative x-ray structure analysis. [9] On the other hand, Steulmann also speculated about the existence of some sort of dipoles in these crystals. All of these ideas proved to be true, and in early 1935, KH2PO4 was found to be ferroelectric with a critical temperature of about 123 K by Busch and Scherrer [10].

This breakthrough discovery in the field of ferroelectricity implied that KH_2PO_4 had become a model substance. Furthermore in 1938 [11], in his thesis, Busch pointed out that the hydrogen bonds between adjacent oxygen atoms might be responsible for ferroelectricity. Giving the first significant molecular theory of ferroelectricity based on the model of hydrogen bonds, Slater addressed the same idea in 1941. [12] Busch then began to investigate $NH_4H_3IO_6$, a material in which hydrogen bonds were also known to exist. Later, this ammonium iodate was determined to be antiferroelectric. Until then, many substances exhibiting hydrogen-bond-ferroelectricity were known and ferroelectricity was no longer a singularity of Rochelle salt.

The Second World War triggered the interest in ferroelectrics, mainly for sonar systems to detect submarines and other military applications. The first man-made perovskite ferroelectric, Barium Titanate ($BaTiO_3$), was discovered in 1942 and 1944 in the United States, Russia and Japan. In the United States, the research was accelerated during the war. At that time, mica was used in most capacitors. However, the supplies of mica to the U.S. from South America were threatened by German U-boats. To produce ceramic materials with enhanced dielectric permittivities, the initial efforts were based on doping studies of TiO_2 with BaO, through which the mixed oxides were fabricated in 1941 (filing date of U.S. Patent No. 2,429,588) [13]. The high permittivities were measured at the Erie Resistor Company, with dielectric constants exceeding 1000, ten times greater than any other ceramic known at that time.

The competition and efforts to understand the nature of the dielectric anomaly in $BaTiO_3$ ceramics continued. In 1945 and 1946, von Hippel (USA) and Goldman (Russia) discovered the ferroelectric switching in these ceramics [14, 15]. This discovery in $BaTiO_3$ ceramics was extremely important. For the first time, the fact that the ferroelectricity could exist in simple oxide materials and was not always associated with hydrogen bonding was demonstrated and proved. $BaTiO_3$ is a member of the perovskite family named after the natural mineral perovskite $CaTiO_3$. The simplicity of the structure of $BaTiO_3$ (Figure 1.1) allowed several investigations to study the structural changes occurring at the phase transitions by undertaking x-ray diffraction studies [16] and enabled the investigation on theoretical backgrounds of these materials.

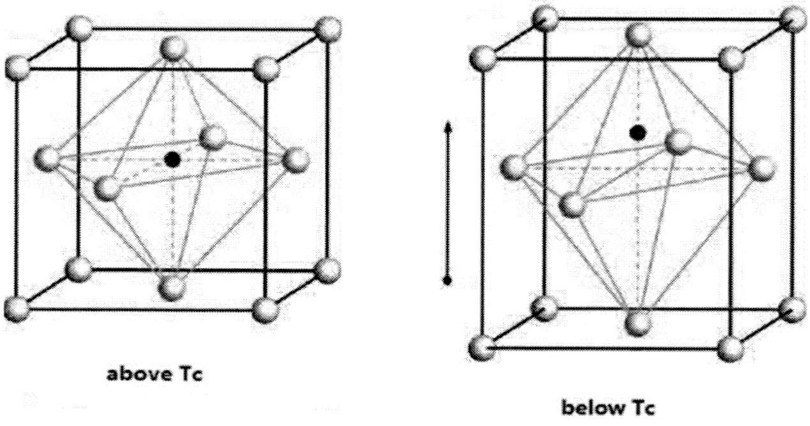

Figure 1.1. The $BaTiO_3$ crystal structure.

In 1951, A.F. Devonshire developed a phenomenological model of ferroelectrics on the basis of the electromechanical, structural and thermal properties of $BaTiO_3$ [17], which built on the preliminary ideas of Landau and Ginzberg, and invoked the point group symmetry and nonlinear elasto-dielectric coupled interactions with electrostriction and piezoelectric coefficients. The resulting Landau-Devonshire-Ginzberg Theory (LDGT) of ferroelectrics is still the most powerful tool for the understanding of ferroelectrics.

At the end of the WWII, public disclosure led to intense interest not only in academic circles, but in practical markets. In particular high capacitance, small volume capacitors in early television and radio circuits, as well as active elements for phonograph pick-ups, accelerometers, and ultrasonic generators, the practical utilization was well developed. For instance, the British Navy used some underwater sonar devices based on the piezoelectric response of $BaTiO_3$ ceramics to detect U-boats, until the end of the last century. Lately, the BaTiO3 piezoelectric properties have been re-used in the form of single crystals and textured ceramics because of environmental concerns with the lead-based perovskites based on the perovskite $PbZrO_3$-$PbTiO_3$ system (PZT).

The $PbZrO_3$-$PbTiO_3$ (PZT) System

The solid solution of the $PbZrO_3$-$PbTiO_3$ system is typically written as $Pb(Zr_xTi_{1-x})O_3$ and is referred to as lead zirconate titanate (PZT), which is the primary ferroelectric materials utilized in the ferroelectric and piezoelectric research fields. PZT has the cubic perovskite structure in the paraelectric phase, and has several ferroelectric phases that are all slight deviations from the cubic structure. The tetragonally distorted perovskite structure is distorted along the polarization direction with Pb atoms at the corners of the tetragon and the Zr/Ti atoms are displaced from the center. While in the cubic perovskite structure, Pb atoms occupy the corners of a cube, with O atoms at the face centers and a Zr or Ti atom at the exact center of the unit cell.

With the different Zr/Ti ratios, the Curie point of PZT varies from 230°C to 490°C. The ferroelectric rhombohedral phase is for high Zr concentrations, and tetragonal phase for high Ti concentrations. In the tetragonal phase, the Zr/Ti atom shifts in a cubic <100> direction and displaces off center towards any one of the six oxygen atoms, and the Pb and O atoms shift as well, is a schematic of the possible orientations of polarization in tetragonal PZT. In the rhombohedral phase, the Zr/Ti atom is shifted in a cubic <111> direction and displaced towards the center of one of the eight faces of the oxygen octahedra. Figure 1.2 shows a schematic of the possible orientations of polarization in rhombohedral PZT. In both cases, the centers of positive and negative charge are offset, generating a spontaneous polarization.

Barium-Strontium Titanate, $(Ba,Sr)TiO_3$ or BST, is the other significant perovskite with ferroelectric properties, in which the A-site of $BaTiO_3$ is doped with Sr which alters the Curie temperature of $BaTiO_3$. Above the Curie temperature, the ferroelectric material becomes paraelectric, and the hysteresis cannot be observed. Below the Curie temperature, the material behaves as a "normal" ferroelectric, and it has a hysteresis like Valasek observed in Rochelle Salt. In terms of tunability (the change of the permittivity with an applied voltage) the paraelectric state offers the opportunity for varactors (variable capacitors) and phase shifters with a linear change of the permittivity in a certain voltage region.

Although BST has been a matter of investigation for over 60 years, this material has problems, such as high growth temperatures (700 – 1000°C) and large leakage current densities, which are some of the major obstacles for these material to be used in dynamic random access memories (DRAM) or microwave dielectric applications. The large leakage current densities limit the use of thinner films and the high growth temperature is detrimental for the integration to semiconductor devices.

To overcome these disadvantages of BST, in recent years a new perovskite material, Lead Strontium Titanate, $(Pb, Sr)TiO_3$ or PST, has been developed and attracted the interest of researchers. Although there are health and environmental concerns with lead-based materials, this material has remarkable advantages compared to BST. It has a comparably low growth temperature (<700°C), relatively low loss, a high tunability (~80%), and is easy to deposit crack-free. Therefore $(Pb,Sr)TiO_3$ is an attractive candidate to replace $(Ba,Sr)TiO_3$ in future applications.

Coming back to Seignette Salt and the term *ferroelectric*, Helen D. Megaw from the United Kingdom said in 1957: "... perhaps the real reason for its [i.e. of the name Seignette electricity] rejection ... is its failure to fit comfortably into the English language. As an adjective, *ferroelectric* is euphonious, while *Seignette-electric* grates on the ear" [18].

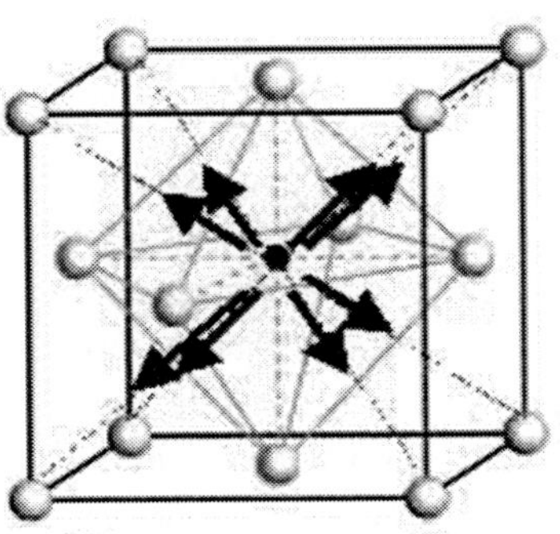

Figure 1.2. The possible orientations of polarization in rhombohedral PZT.

1.2. Basic Knowledge of Ferroelectricity

1.2.1. Simple Knowledge of Piezoelectric Effect

Since Pierre and Jacques Curie discovered piezoelectricity in 1880, [1] the piezoelectric effect was found to occur in most classes of non-centrosymmetric crystals (class 432 cubic is the lone exception). The electrical polarization proportional to an applied mechanical stress also occurs in these non-centrosymmetric and piezoelectric crystal classes, which is called the direct piezoelectric effect. [19] On the other hand, the converse piezoelectric effect also occurs in all piezoelectric materials, in which a geometric strain is proportional to an applied electric field. Ignoring tensor components, the mathematical expressions for the direct and converse piezoelectric effects are given by the following equations:

$P = d \cdot T$ (for direct Effect), and

$S = d \cdot E$ (for converse Effect),

where the proportionality coefficient is the piezoelectric coefficient (d) and the variables are polarization (P), stress (T), mechanical strain (S) and electric field (E).

The fact that the ceramic barium titanate could generate piezoelectricity by the application of a suitable electric field, was first reported independently by Gray and Roberts. ABulk piezoelectricity was long believed to be unique to a certain class of single crystals, instead of bulk ceramics with their isotropic nature that made the existence of piezoelectric in ceramics unthinkable. [20, 21] It is well known that piezoelectricity in a isotropic ceramic depends on ferroelectricity. By application of a suitable electric field, ferroelectric crystals with the non-centrosymmetric structures exhibit a spontaneous electric polarization that can be reoriented between two or more states. It is the existence of multiple equivalent thermodynamic states that lead to the ability to reorient these ferroelectric materials.

The so-called local minima is directly originated from the symmetry of the crystal. Ferroelectric materials are low symmetry (polar) structures, and exhibit several degenerate states that are subsets of a higher symmetry (polar or non-polar) phase. The symmetry of each degenerate state is linked to the symmetry of the "prototype" state, and the high symmetry phase may be either a real or hypothetical phase. The symmetry of the ferroelectric phase can be defined by the symmetry of the high temperature phase and the orientation of spontaneous polarization (PS) based on the initial phase and according to the Curie Principle. A number of possible ferroelectric phases may exist according to a given prototype phase, which are each subsets of the symmetry of the high temperature phase.6 In addition, the equivalent state numbers, are defined by the symmetry of the crystallographic axes along which polarization is developed in the prototype phase. While the ceramics with randomly oriented microstructures are advantageous in having a large of number equivalent polar directions, to increase the ability to induce a more uniform preferred orientation in the bulk ceramic. [22, 23]

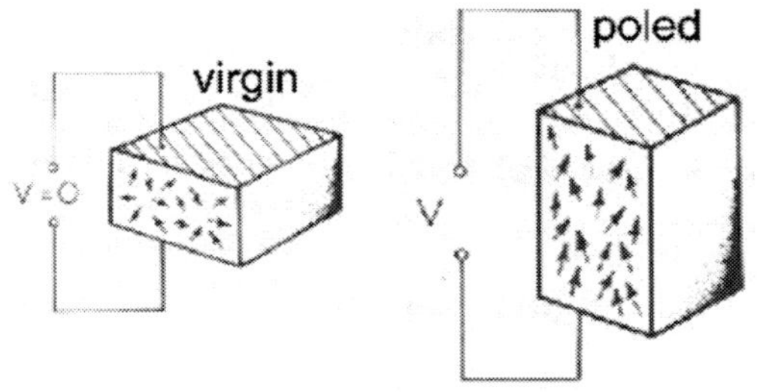

Figure 1.3. The change in the polarization state and shape (exaggerated) between a virgin and a poled ceramic sample.

The so-called poling is defined by the process of orienting the polarization (Figure 1. 3). By the application of a critical field (EC), the polarization vectors can be oriented in the allowed crystallographic directions nearest the applied field. When removed off the electric field there exists a net remanent polarization (PR). Analogous to the magnetization-magnetic field in ferromagnetic materials, the application of a bipolar field can produce switching between differently polarized states, leading to a polarization-electric field hysteresis loop in ferroelectric materials.

1.2.2. Class of Ferroelectric Crystal Structure

Piezoelectric materials become electrically polarized when they are strained (direct piezoelectric effect), and an electric field applied to the materials causes the crystal to strain (converse piezoelectric effect). It is well known the centrosymmetric crystal does not possess polar properties and cannot possess piezoelectric function. There are 21 crystal classes that lack a center of symmetry, in which 20 are piezoelectric (except Class 432, which is non-piezoelectric non-centrosymmetric class). [24]

Among the piezoelectric classes, ten classes are pyroelectric, in which there is a spontaneous polarization, i.e., an electric polarization that exists in the absence of an applied electric field or stress. Ferroelectric materials are a subset of pyroelectric materials in that they have a switchable spontaneous polarization. This polarization originates from a displacement between the centers of positive and negative charge in the crystal unit cell and changes when the material is heated uniformly causing a charge to move on the surface.

1.2.3. Definition of Ferroelectricity [24]

In describing phenomena specific to ferroelectric materials, we know that ferroelectric materials are a subset of pyroelectrics and in turn are a subset of piezoelectrics (ferroelectrics belong to one of the 10 pyroelectric crystal classes). [24] Ferroelectrics possess a spontaneous electric dipole moment, develop charge on their surfaces when they subjected to stress and change shape under an electric field. The main ferroelectric characteristic in these materials that is different from pyroelectrics, is the presence of a switchable spontaneous polarization. Generally, ferroelectric materials exhibit ferroelectric behavior below a certain temperature and paraelectric behavior above this temperature. This temperature is called the Curie point

(Tc), indicating the occurrence of structural phase transition from a high temperature nonferroelectric paraelectric phase to a low-temperature ferroelectric phase.

1.2.4. Ferroelectric Domains [24]

There exist regions of uniform polarization direction in ferroelectric materials, although the magnitude and direction of polarization are typically not constant in the materials. These regions are called ferroelectric domains (similar to the domains of ferromagnetism), and the boundaries between them are called domain walls. The formation of domains is governed by energetic considerations; i.e. the minimization of the total crystal energy. [25] These domains are largely responsible for the physical characteristics of ferroelectrics.

Generally, a ferroelectric crystal exhibits uniform spontaneous polarization, with charges induced on the external surface that create a depolarizing field leading to the formation of domains. These domains are also responsible for reducing the depolarizing field. When the energy stored in the domain walls offset the reduction in energy of the depolarizing field, an equilibrium is reached. When the ferroelectric material is cooled through the Curie point, the domains form through the process of domain nucleation, which is one of the most important factors governing domain formation. The domain formation is also obviously affected by strain with imperfections such as dislocations and vacancies, interfacial effects, and finite conductivity. However, for minimizing the domain wall charge, domains form to minimize the polarization discontinuity at domain walls.

There are two types of domain walls, classified as: 180° domain walls (those with antiparallel polarizations on either side of the wall), and non-180° walls. In the ideal case, the tetragonal ferroelectric materials exhibit the 90° wall (the only non-180° domain wall) and the polarization direction rotates by 90° from one domain to the other. When the normal components of polarization are of discontinuity, a surface charge always exists. Since 180° domain walls have no polarization component normal to the wall, there is no surface charge. However, for the case of the 90° domain, the domain walls are at approximately 45° to the polarization directions on either side of the wall. Therefore, the normal component is nearly continuous across the 90° domain wall. Based on the fact that the total crystal energy is to be minimized, the non-through domains that do not extend the entire thickness of the materials can be formed with charged head-to-head or tail-to-tail domain walls.

There are many methods and techniques that can observe and image the ferroelectric domains, [26-42] including polarized optical microscopy, electron microscopy, electron emission, etching, and various scanning probe techniques.

1.2.5. Polarization Hysteresis [24]

The most important characteristic for ferroelectrics is that these materials exhibit polarization hysteresis. This can be considered a resistance to change from a previously induced condition. For instance, a uniform ferroelectric crystal with zero net polarization can be considered as a result of having an equal number of antiparallel domains. By application of a voltage and an electric field to the crystal, the polarization increases linearly as a linear dielectric does. (origin to A in Figure 1.4)

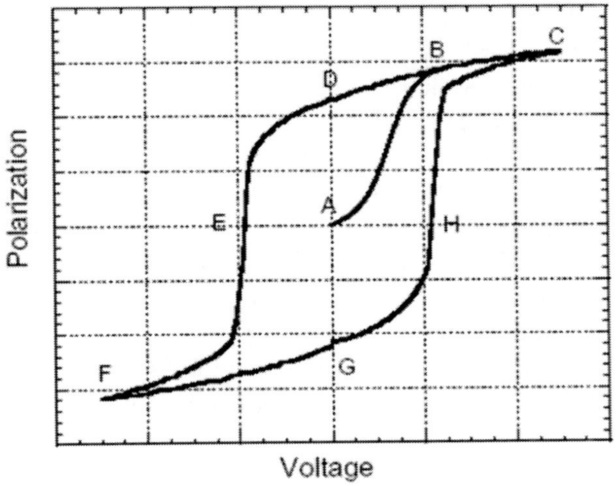

Figure 1.4. A typical polarization hysteresis (P-V) loop.

By increasing the electric fields, the polarization in the domains which are parallel to the applied field remains unchanged; but the domains with polarization that is antiparallel to the electric field experience a torque.

As the electric field is sufficiently strong so that is able to switch the direction of the antiparallel domains, the net polarization increases rapidly (A to B), until all of the domains become parallel to the field (point C). By decreasing the electric field (C to D) , the polarization does not decrease to zero; in fact, the switched domains tend to remain polarized which are parallel to the electric field, even though the electric field is reduced to zero. This is the phenomenon that is called positive remenant polarization. To reach again to zero net polarization, a field has to be applied in the opposite direction by applying a negative voltage. At this moment, antiparallel domains switch to the point of zero net polarization (point E), until the polarization completely becomes parallel to the electric field (point F). When the electric field is removed, a negative remenant polarization (point G) can be retained. By increasing the electric field from G to H, the polarization decreases until the net polarization is zero, as followed by an increase in polarization until the crystal is uniformly polarized parallel to the applied field. The points E and H are respectively represented as the negative and positive coercive voltages.

The traditional method for measuring the polarization hysteresis is that P-E or P-V measure the charge flow to a ferroelectric capacitor. This method can measure a change in polarization. But by these traditional P-E or P-V measurements, a fixed polarization within a ferroelectric thin film cannot be given. The recently developed d_{33}-V measurements related to PFM have an advantage in doing this measurement, due to their ability to built-in fields and sensitivity to fixed polarization.

1.2.6. Imprint

Typically, imprint is manifested as a voltage shift in P-V or d_{33}-V loopsis, [24] which is defined as the tendency of one polarization state to become more stable than the opposite one.

In general, imprint can be attributed to interface charge entrapment. The voltage shift can be ascribed to an asymmetric distribution of charged defects or the presence of defect dipoles. For instance, in ferroelectric capacitors, trapped charges at the ends of polarization vectors tend to stabilize one state, which can be influenced by thermal and optical processes. By heating, ferroelectrics can become imprinted. However, heating is not necessary to generate a thermal imprint. If a ferroelectric capacitor is left at a positive remenant polarization, a P-V (or d_{33}-V) loop will shift towards the left. If the capacitor is left at a negative remenant polarization, the loop will shift to the right. Another imprint of optical imprint can be generated by UV illumination. The illumination excites free carries or charges in a material, making the material become imprinted. In non-volatile ferroelectric random access memory (NV-FRAM) devices, the imprint became an important failure mechanism. A voltage shift in the hysteresis loop can lead to both an increase in the coercive voltage (possibly causing a write failure) and a decrease in remenant polarization (possibly causing a read failure).

1.2.7. Fatigue [44-65]

Like imprint, fatigue is an important failure mechanism in NV-FRAM and leads to read failure since the remenant polarization is decreased. Fatigue is defined as the loss of switchable polarization as a result of repeated switching (electrical stress resulting from field cycling) and it manifests itself as a reduction in polarization (piezoresponse) in P-V (d_{33}-V) loops. Fatigue has been studied since the 1950's, although the origin of this phenomenon is not fully understood.

There are various fatigue models. In analyzing experimental data from different groups, the mechanisms for the reduction of switchable polarization had been summarized to be the reduction of the effective electrical field through the presence of a nonswitching "passive" layer or a modification of the switching process, which is attributed to either wall pinning or nucleation inhibition. Both the formation of a passive layer and the switching process modification involve the density of defects.

On the other hand, the fatigue mechanism is also ascribed to the inhibition of domain wall motion induced by charged carriers or oxygen vacancies (ionic defects). The fatigue theories based on polarization are generally related to the inhibition that occurs in regions, such as the area of an electrode-ferroelectric interface. Since the oxygen vacancies are the most mobile and numerous of the defects, they are considered to be the most likely ionic defect. However, it was concluded that it was the electron injection that plays the predominant role and drives the force of fatigue.

First, in the case for that the inhibition occurs in the bulk of the material, the fatigue is ascribed to the bulk-locking mechanism. The domain walls become locked or immobile and stop switching. Since the domain walls with a polarization discontinuity have a net surface charge, the mobile charges (electronic or ionic) in the bulk that are redistributed during the repeated switching, will be attracted to the domain wall surface charge in order to compensate for these charged walls. In this process, to stabilize the domain polarization and suppress the switchable polarization, an internal field will be created (fatigue is sometimes considered as a form of imprint). In addition, oxygen vacancies are also created during the field cycling, which result in an internal bias. However, the source of the internal bias is not necessarily due to oxygen vacancies; UV illumination near the coercive field generates electron-hole pairs, which also result in polarization fatigue in ferroelectric capacitors. Therefore, bulk-locking fatigue is probably due to electronic charge. (not ionic charge)

In the case of ferroelectric capacitors, the ferroelectric-electrode interface must be taken into account to understand the polarization fatigue. This causes the widely used PZT capacitors with Pt electrodes to fatigue severely, while the PZT capacitors with conducting oxide electrodes (e.g. RuO2, IrO2, etc.) have superior fatigue properties. 70, 72-76 suggest that the oxygen vacancies are located or trapped at the ferroelectric-electrode interface. These trapped charges can generate an internal field that hinders domain switching. This internal field as well as the fatigue can be reduced, since with oxide electrodes, the oxygen vacancies could be compensated for by oxygen from the electrode. To determine whether the fatigue is due to oxygen vacancies in the bulk or at an interface, it is necessary to know the distribution of oxygen vacancies.

In continuing the studies with the interface model of fatigue, it was suggested that the domain nucleation and switching can be inhibited near the electrode-ferroelectric interface. The switchable polarization is possibly due to the inhibition of nucleation or to the locking of opposite polarization nucleation sites at electrode-ferroelectric interfaces. The results that the hysteresis loops can offset vertically in fatigued films, and from this asymmetry one can deduce whether nucleation is suppressed more in one direction than the other.

In addition to the two fatigue models, several models were proposed for the combinations of bulk and interface theory. Du et al. proposed a model that pointed out that both oxygen vacancies and electronic charge carriers were necessary to induce fatigue. Since the presence of electrons facilitates the formation of oxygen vacancies, a space charge layer can form near the electrode (interface nucleation inhibition) and make domain walls immobile within the film (bulk-locking). In addition, a fatigue model that suggests oxygen vacancies are trapped at both the ferroelectric-electrode interface and domain and grain boundaries in the bulk has also been proposed.

1.2.8. Leakage Measurement [66]

The leakage current measurement performs an I vs. V measurement on the sample, which allows the detection of high leakage currents. The leakage current characteristic affects the shape of the hysteresis loop, especially at high voltages and low frequencies.

There are two types of leakage current: dc leakage current and ac leakage current. Dc leakage current usually applies only to end-product equipment and not to power supplies. Ac leakage current is induced by combining the capacitance and dc resistance between a voltage source (ac line) and the grounded conductive parts of the equipment. The leakage caused by the dc resistance usually is insignificant compared to the ac impedance of various parallel capacitances, which may be intentional or unintentional. Some examples of unintentional capacitances are spacings on printed wiring boards, insulations between semiconductors and grounded heatsinks, and the primary-to-secondary capacitance of isolating transformers within the power supply.

It is important to understand the leakage current. One of the interesting subjects in the field of ferroelectric devices is the development of high-density non-volatile ferroelectric memories (NvFRAM). Many ferroelectrics such as $Pb(Zr, Ti)O_3$ (PZT) ferroelectrics, are promising NvFRAM materials. These materials had been tentatively studied in order to improve their properties including polarization field (P-E) characteristics, degradations such as fatigue, retention and imprint, and leakage current. Extensive research in leakage current has been devoted to understanding or improving/removing the P-E hysteresis characteristics and their degradations.

Although a variety of physical models describing the leakage current behaviors of ferroelectrics have been proposed, the comprehensive understanding of them is not yet fully established. For instance, the proposed mechanisms of leakage conduction for PZT are bulk ionic conduction, Schottky emission, space-charge-limited conduction (SCLC), and mixed Schottky and Pool-Frankel emission. Many intrinsic and extrinsic factors devoted to a variety of experimental results and interpretations concerning the leakage conduction of ferroelectric thin film capacitors, that included the intrinsic effects due to different measuring techniques and the extrinsic effects due to different electrode/ferroelectrics interfaces. In these interfaces, the effect of the inter-diffusion layers, oxygen depletion layers and interface traps cannot be neglected. In many other cases, the conclusions that are based only on analyzing the shapes of the J-V curves can be ambiguous, and the different conduction mechanisms may possibly result in different or similar J-V curves. [67-70]

The main problem is that the leakage current in ferroelectric films is hard to measure directly. The current measure for ferroelectric thin film capacitors is the charging current (J_C) of the test capacitors. This current includes the switching current (J_S), the polarization current (J_P), and the true leakage current (J_L). Additionally, a current due to the aging effect (J_A) can also influence the measured charging current. Therefore, J_C can be described as follows:

$$J_C = J_S + J_A + J_P + J_L$$

Since J_S contributes during the few microseconds in the fully switching state and J_A appears normally at high fields and/or at high temperature and can be discriminated from the J-t curves, for a given time interval, temperature, and measuring time, the measured J_C is dominated by J_P and J_L:

$$J_C \cong J_P + J_L$$

However, the current response in ferroelectric films includes a time-dependent polarization current J_P (t) of dielectric relaxation. This time-dependent current depends strongly on the measuring technique and the prehistory treatment for samples, such as the applied poling electric field. Recently, Stolichnov and Tagantsev [71] proposed that the true leakage current of PZT thin films was time dependent because of the influence of injected charge entrapment during the measurement.

1.3. DIELECTRIC BEHAVIORS [72, 73]

1.3.1. Dielectric

Ferroelectrics can also be considered dielectric materials. A **dielectric** is an electrical insulator that can be polarized by an applied electric field. When a dielectric is applied on an electric field, electric charges do not flow through the material as they do in a conductor, but only slightly shift from their average equilibrium positions. This behavior causes dielectric polarization in the materials, resulting in the positive charges will be displaced toward the field and negative charges shift in the opposite direction, and creating an internal electric field which reduces the overall field within the dielectric materials. In a dielectric that is composed of weakly bonded molecules, the molecules will become polarized and reorient to make their symmetry axis align to the field.

A common example of a dielectric is the electrically insulating material between the metallic plates of a capacitor. Although the dielectric exhibits low electrical conduction, the dielectric is typically used to describe materials with a high polarizability and is used to indicate the energy storage capacity of the material (by means of polarization). The polarization of the dielectric by the applied electric field increases the capacitor's surface charge. Therefore, the study of dielectric properties is concerned with the storage and dissipation of electric and magnetic energy in materials.

This characteristic in dielectrics is also expressed by a number called the dielectric constant. It is important to explain various phenomena in electronics, optics, and solid-state.

Dielectric Polarization

Basic Atomic Model

Since a material is made up of atoms, a classical approach to the dielectric model of atomic model was proposed for understanding the dielectric polarization. Each atom consists of a cloud of negative charge (electrons) bound to and surrounding a positive point charge at its center. In the presence of an electric field the charge cloud is distorted, which can be reduced to a simple dipole, that is characterized by its dipole moment. It is the relationship between the electric field and the dipole moment that gives rise to the behavior of the dielectric.

The relationship between the electric field **E** and the dipole moment **M** gives rise to the behavior of the dielectric, which, for a given material, can be characterized by the function **F** defined by the equation:

$$\mathbf{M} = \mathbf{F}(\mathbf{E}).$$

When both the type of electric field and the type of material have been defined, one then chooses the simplest function F that correctly predicts the phenomena of interest.

For dielectrics, when the electric field is removed, the atom returns to its original state. The time required to complete this process is called relaxation time, obeying an exponential decay rule. This is the physical essence for this model. The behavior of the dielectric now depends on the situation. To accurately describe the behavior, There needs to be a more complicated situation and richer model.

In addition, to study the dielectric polarization, some of the following important questions must be answered:

(1) Is the electric field constant or does it vary with time? If the electric field does vary, at what rate?
(2) What are the characteristics of the material (isotropy or homogeneous)? Are there any boundaries/interfaces that have to be taken into account?
(3) Is the system linear or do nonlinearities have to be taken into account?

Dipolar Polarization

Dipolar polarization is a polarization that is inherent to polar (orientation polarization), which can be induced in any molecule in which the asymmetric distortion of the nuclei is possible (distortion polarization). Orientation polarization results from a permanent dipole, which retains polarization in the absence of an external electric field. For instance, in the water molecule, the polarization arising from the 104.45° angle between the asymmetric bonds between oxygen and hydrogen atoms. These dipoles form an assembled macroscopic polarization.

When an external electric field is applied, the distance between charges remains constant in orientation polarization. But the polarization itself rotates under the applied field, which occurs on a timescale depending on the torque and the surrounding local viscosity of the molecules. At the lowest frequency, since the rotation is not instantaneous, the dipolar polarizations lose the response to electric fields.

A molecule will be bent and stretched by the electric field when this external electric field is applied in the infrared, and the molecular moment changes in response. In this case, the molecular vibration frequency is approximately the inverse of the time taken for the molecule to bend, and the distortion polarization will disappear above the infrared.

Ionic Polarization

In ionic crystals, the polarization is generally ionic polarization, which is caused by the displacements between positive and negative ions. Generally, when lattice vibrations or molecular vibrations induce relative displacements of the atoms, the centers of positive and negative charges might be in different locations, which are affected by the symmetry of the displacements. When the centers don't correspond to this case, polarizations may arise in molecules or crystals.

Ionic polarization causes ferroelectric transition as well as dipolar polarization. The transition, which is caused by the order of the directional orientations of permanent dipoles along a particular direction, is called order-disorder phase transition. The transition which is caused by ionic polarizations in crystals is called displacive phase transition.

Dielectric Dispersion

Since there is a delay between changes in polarization and changes in an electric field, and the permittivity of the dielectric is a complicated function of frequency of the electric field, dielectric dispersion in physics is the dependence of the permittivity of a dielectric material on the frequency of an applied electric field. To explore the phenomena of dielectric dispersion is very important for the application of dielectric materials and the analysis of polarization systems.

Generally, the phenomenon known as material dispersion exhibits a frequency-dependent response of a medium for wave propagation. With different frequencies, (1) the dipolar polarization is not possible to follow the electric field changes in the microwave region around 10^{10} Hz; (2)ionic polarization and molecular distortion polarization will lose the response to the electric field, in the infrared or far-infrared region around 10^{13} Hz. (3) in the ultraviolet region around 10^{15} Hz, electronic polarization loses its response.

For the frequency region above ultraviolet, permittivity approaches the constant ε_0 in every substance, where ε_0 is the permittivity of the free space. Since permittivity indicates the strength of the relation between an electric field and polarization, if a polarization process loses its response, permittivity decreases.

Dielectric Relaxation

Dielectric relaxation indicates the momentary delay in the dielectric constant of a material. This is generally caused by the delay in the polarization with respect to a changing electric field. B (for inside, capacitors or between two large conducting surfaces). This kind of relaxation in changing electric fields could be analogous to the hysteresis in changing magnetic fields (for inductors or transformers). Relaxation, in general, is also a delay in the response of a linear system, and therefore dielectric relaxation can be measured relative to the expected linear steady-state (equilibrium) dielectric values. The time delay between electrical field and polarization implies an irreversible degradation of free energy(G).

In physics, dielectric relaxation refers to the relaxation response of a dielectric to an external electric field within microwave frequencies, which is often described in terms of permittivity as a function of frequency, and can be described by the Debye equation for ideal systems. On the other hand, the distortion related to ionic and electronic polarization shows behavior of the resonance or oscillator type. The character of the distortion process depends on the structure, composition, and factors of the sample.

The number of possible wavelengths of emitted radiation due to dielectric relaxation can be equated using Hemmings' first law:

$$n = \frac{l^2 - l}{2}$$

where n is the number of different possible wavelengths of emitted radiation l is the number of energy levels (including ground level).

Debye Relaxation

Debye relaxation can be considered the dielectric relaxation response of an ideal and noninteracting population of dipoles to an alternating external electric field. It is usually expressed in the complex permittivity ε of a medium as a function of the field's frequency ω:

$$\hat{\varepsilon}(\omega) = \varepsilon_{\infty} + \frac{\Delta\varepsilon}{1 + i\omega\tau},$$

where first term is the permittivity at the high frequency limit, where is the static, low-frequency permittivity, and is the characteristic relaxation of the medium.

There are various equations of the Debye relaxation, such as Cole–Cole equation, Cole–Davidson equation, Havriliak–Negami relaxation, and Kohlrausch–Williams–Watts function (Fourier transform of stretched exponential function).

Applications

Solid dielectric materials with high permittivity can be used in commercially manufactured capacitors as the typically intervening medium between the stored positive and negative charges, which are often referred to as capacitor dielectric. [6]

The most important advantage for using such a dielectric material is that it prevents the conducting plates on which the charges are stored from coming into direct electrical contact. More significantly at a given voltage, a high permittivity allows a greater charge to be stored, which can be treated as the case of a linear dielectric with permittivity ε and thickness d between two conducting plates with uniform charge density σ_ε. The charge density in this case is as following:

$$\sigma_\varepsilon = \varepsilon \frac{V}{d}$$

and the capacitance per unit area is:

$$c = \frac{\sigma_\varepsilon}{V} = \frac{\varepsilon}{d}$$

Based on this, a larger ε leads to greater charge stored and greater capacitance.

The capacitor dielectric materials are also chosen for they are resistant to ionization, which allows the capacitor to operate at higher voltages before the insulating dielectric ionizes, and begins to allow undesirable current.

Dielectric Resonator

The dielectric resonator oscillator (DRO) can be considered an electronic component exhibiting resonance for a narrow range of microwave frequencies, which consists of a "puck" of ceramic that has a large dielectric constant and a low dissipation factor. The dielectric resonators are often used to provide a frequency reference in an oscillator circuit. An unshielded dielectric resonator can also be used as a Dielectric Resonator Antenna (DRA)

1.3.2. Impedance Spectroscopy

In the studies for advanced electroceramics, impedance Spectroscopy (IS) has become a powerful, nondestructive technique that has been widely used to characterize their electrical properties. To relate the impedance/dielectric spectroscopy (I/DS) data to the different electro-active regions in the microstructure (grain core, grain boundary), a suitable electrical analogy is needed. An alternative technique for measuring electrical properties is impedance spectroscopy.

The **AC** impedance measurements can be performed over a wide range of frequencies, and the different regions of the material can be characterized according to their electrical relaxation times or time constants. Impedance spectroscopy is relatively easy to use and can be applicable to a wide variety of materials and problems. This technique has recently undergone major developments with the availability of automatic equipment, which is capable of spanning many decades of frequency in a single sweep. It can take several minutes to obtain a single reading, in comparison with the use of null methods, which can take several days for balancing ***R*** and *C* (or ***L***) circuits.

Impedance spectroscopy is useful for the characterization of electroceramics and it enables the overall electrical properties of a material to be separated into their component parts. It is rapidly becoming an essential technique in the development of new and improved materials.

Fundamentals for Impedance Spectroscopy

The impedance of a sample can be measured over a wide range of frequencies, typically 10^2 to 10^7 Hz, by the impedance spectroscopy. Impedances usually have both resistive and reactive (capacitive/inductive) components, which have to be determined. This measurement can be achieved in various ways: one is to apply an alternating voltage across a sample and a standard resistor which are in series, then measure the in- and out-of-phase components of the voltage across the sample. Through dividing these components by the magnitude of the current, the resistive and reactive components of the impedance can be given. The measurements can be repeated as a step-wise function of frequency. [74]

Usually by placing a resistance and a capacitance in parallel, different regions of a ceramic sample can be characterized. The characteristic relaxation time or time constant of each 'parallel *RC* element' can be given by the product of ***R*** and *C*

$$\tau = RC \quad (1)$$

$$\omega_{max} RC = 1 \quad (2)$$

From the impedance spectrum, it is possible to identify different *RC* elements and assign them to the appropriate regions of the sample. The values of the individual *R* and *C* components may then be quantified. The following listed some practical examples of data and their interpretation.

A common type impedance spectrum for electroceramics exhibits two distinct features, which are ascribed to intragrain, or bulk and intergrain, or grain boundary regions. After having obtained values for the *R* and *C* components, the next stage is to assign them to regions of the sample. The assignment is based on the magnitudes of the capacitances. For a parallel plate capacitor with area A, separation l between the plates and a medium of permittivity between the plates, the capacitance is given as following:

$$C = \varepsilon' e_0 \frac{A}{l} \tag{3}$$

where ε_0 is the permittivity of free space. Figure 1.5 shows high frequency semicircles, associating resistance that is attributed to the bulk properties of the sample.

To assign the second semicircle to a feature of the ceramic, it is essential to have an idealized ceramic with grains and grain boundaries, and consider the factors which control the grain boundary impedance magnitude. The idealized case can be illuminated in Equation 4, which arises from the inverse relation between thickness and capacitance given by Equation 3.

A wide variety of ceramic microstructures occur in practice and it is found that the grain boundary capacitance usually lies in the range 10^{-11} to 10-8 F. The higher capacitances occur in materials that are well-sintered, with narrow intergranular regions. The lower capacitances are often found with poorly sintered samples that contain constriction resistances or narrow contacting. For the case in Figure 1, the grain boundary capacitance is quite large, indicating a well-sintered sample. Nevertheless, the overall resistance of these narrow grain boundaries is 2-3 times greater than the resistance of the grains.

$$\frac{C_b}{C_{gb}} = \frac{l_2}{l_1} \tag{4}$$

Results such as those shown in Figure 1 and Table 1.1 are useful for several reasons: to indicate whether the overall resistance of a material is dominated by bulk or grain boundary components; to assess the quality and electrical homogeneity of an electroceramic, since there is generally a link between sintering/microstructure and AC response - to measure the values of the component resistances and capacitances.

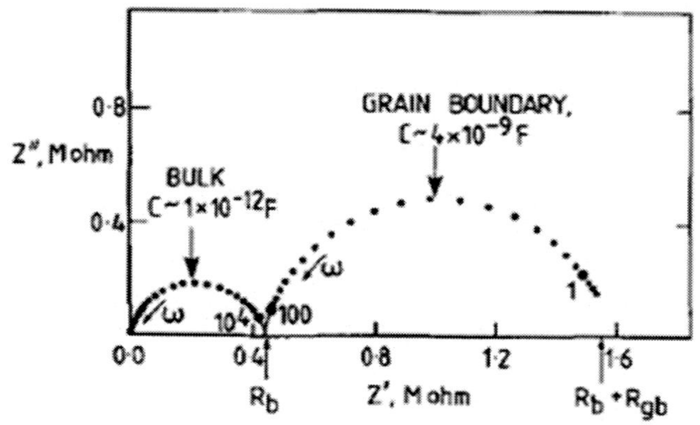

Figure 1.5. A typical impedance spectrum.

Table 1.1. Capacitance values and interpretation

Capacitance [F}	Phenomenon Responsible
10^{-12}	bulk
10^{-11}	minor, second phase
10^{-11} - 10^{-8}	grain boundary
10^{-10} - 10^{-9}	bulk ferroelectric
10^{-9} – 10^{-7}	surface layer
10^{-7} – 10^{-5}	sample-electrode interface
10^{-4}	electrochemical reactions

1.4. PIEZOELECTRIC BEHAVIORS

The piezoelectric effect was discovered in 1880 by the Curie brothers. When a mechanical stress was applied on crystals such as quartz and Rochelle salt, electrical charges appeared, which were proportional to the stress. The first important applications for piezoelectric effect were piezoelectric ultrasonic transducers and quartz clocks (swinging quartz for standards of frequency).

Introduction: The Piezoelectric Effect

In physics, the piezoelectric effect can be described as the relationship between electrostatics and mechanics. The definition: piezoelectric effect describes the relation between a mechanical stress and an electrical voltage in solids. It is reversible: an applied mechanical stress will generate a voltage and an applied voltage will change the shape of the solid by a small amount (up to a 4% change in volume).

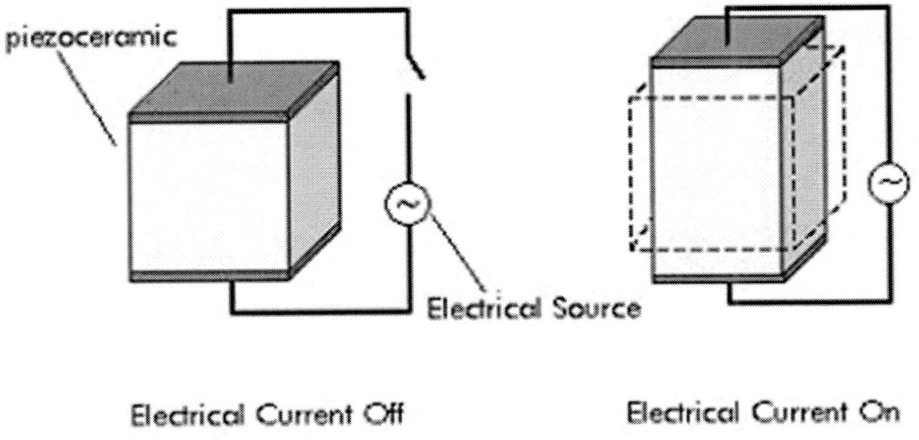

Figure 1.6. piezoelectric effect.

Piezoelectric Materials

Piezoelectric materials can be divided into two main groups: crystals and ceramics. The piezoelectric effect occurs only in non-conductive materials. The most well-known piezoelectric material is quartz (SiO2) and PZT ceramics.

What Is Piezoelectricity?

By squeezing certain crystals (such as quartz) you can make electricity flow through them. The reverse is usually true as well: if you make electricity go through the same crystals, they will vibrate back and forth. A formal definition for piezoelectricity:

Piezoelectric effect (Piezoelectricity) is the appearance of an electrical potential (a voltage) across the sides of a crystal when you subject it to mechanical stress (by squeezing or pressing it).

In principle, the piezoelectric crystal will have a positive charge on one face and a negative charge on the opposite face; the current flows if we connect the two faces together to make a circuit. In the reverse piezoelectric effect, when a voltage is applied across the opposite faces of a crystal, it becomes mechanically stressed (deformed in shape).

What Causes Piezoelectricity?

Normally, piezoelectric crystals are electrically neutral, and the atoms inside them may not be arranged symmetrically. But their electrical charges are perfectly balanced: a positive charge in one place cancels out a negative charge nearby. However, if a piezoelectric crystal is squeezed or stretched , the structure will be deformed, pushing some of the atoms closer together or apart away, upsetting the balance of positive and negative, and causing net electrical charges to appear. This effect carries through the whole structure so that net positive and negative charges appear on opposite, outer faces of the crystal.

The reverse-piezoelectric effect occurs in the opposite way. When a voltage is put across a piezoelectric crystal, the atoms inside it will be subjected to "electrical pressure" and have to move to rebalance themselves—and that's what causes piezoelectric crystals to deform (slightly change shape) when you apply a voltage across them.

Piezoelectric Measurements

Several different ways can be used to measure the piezoelectric constant, and the measurements by the converse and direct piezoelectric effect ideally lead to the same piezoelectric constant.

By employing the piezoelectric equation for the electric flux density (D_i), the direct piezoelectric effect can be utilized to measure piezoelectric constants. With a constant electric field, the change in electric flux density is given by

$$\Delta D_i = d_{ij} \Delta T_j \, .$$

As the direct piezoelectric effect is used to measure piezoelectric constants, under the condition of a constant electric field, a stress is applied to a piezoelectric capacitor, and the total charge flow between the electrodes is then measured. The piezoelectric constant can be determined through D_i, which is also the free surface charge density. For measurement of d_{33}, a stress (T_3) is applied in the direction of the electrodes perpendicular to the **x**-axis. For a d_{31} measurement, stress can be generated via substrate bending. The main difficulties with this type of measurement are applying only the desired stress (T3 or T1), eliminating charge flow due to pyroelectric effects, and measuring charge before it begins to leak off, avoiding the result of a lower measured piezoelectric constant. [75-80]

For the converse effect to be used to measure piezoelectric constants, it requires a very sensitive measurement technique, since small displacements must be measured. When an electric field is applied to a piezoelectric material, the change in strain or the resulting piezoelectric displacement can be measured. Evidently, unless high voltages are applied to the piezoelectric material, a measurement technique with sub-Angstrom vertical resolution is necessary to measure the converse piezoelectric effect.

There are two primary types of techniques for measuring the converse piezoelectric effect: scanning probe based techniques and interferometric techniques. PFM (piezoresponse force microscopy) focus on the measurement of piezoelectrically induced strain, and other scanning probe techniques are used to measure induced strain generally employ scanning tunneling microscopy (STM). PFM and STM have the advantage over interferometric techniques in that they are able to measure the piezoelectric effect of small regions of a sample, depending on the tip radius. Moreover, with PFM and STM, the tip can be rastered to generate piezoelectric response images. The vertical resolution of AFM and STM measurements is approximately 10^{-1}- 10^{-2} Å and can be improved using a standard phase sensitive lock-in technique. Interferometric measurements have better vertical resolution (detection limit of piezoelectric deformation) than the scanning probe techniques (10^{-2} – 10^{-4}Å versus 10^{-1} - 10^{-2}Å), which are macroscopic in nature producing an average over an area roughly the size of the laser beam. With double beam interferometry, an AC field is applied to a piezoelectric capacitor, which makes the capacitor expand and contract at the AC modulation frequency. A laser beam can be split into a reference beam and a measurement beam, which is reflected from both electrodes. The piezoelectric displacements result in a change in optical path length for the second beam, which will result in a change in intensity at a photodiode due to interference with the reference beam. [81-88]

Mechanical and Electrical Quality Factors (QM and QE) [89-91]

QM and QE are the mechanical and electrical quality factors, which are important in determining the power handling ability and bandwidth of piezoelectric devices. The quality factors Q are taken to be the inverse of the mechanical and electric loss tangents respectively. High QE is important in off resonance devices, where heat generation is caused mainly by dielectric loss. The electrical energy loss is related to the area within the polarization–E-field hysteresis loop, where ω_e is the hysteresis area in a full cycle and Ue is the energy stored in a quarter of a cycle.

$$\tan \delta'_E = \left(\frac{1}{2\pi}\right)\left(\frac{w_e}{U_e}\right) \tag{2.6}$$

Strain vs. Stress curves are analogous relationships that can be made for the mechanical loss for short-circuit conditions. For high power applications, both high QE and QM are desirable. In general, the QM is directly proportional to QE. The empirical observations of Gerthsen et al. and later the extrinsic model of Artl et al. have linked QM and QE.

1.5. Pyroelectricity Effect [92-96]

The pyroelectric effect for the ferroelectric materials indicates the development of electric dipoles in certain crystalline materials as a result of temperature changes.

Pyroelectricity is the ability of certain materials to generate a temporary voltage when they are heated or cooled. The change in temperature modifies the positions of the atoms slightly within the crystal structure and changes the polarization of the material. This polarization change gives rise to a voltage across the crystal. If the temperature stays constant at its new value, the pyroelectric voltage gradually disappears due to leakage (the leakage can be due to electrons moving through the crystal, ions moving through the air, etc.).

Pyroelectricity is different from thermoelectricity: in a typical demonstration of pyroelectricity, the whole crystal is changed from one temperature to another, and the result is a temporary voltage across the crystal. In a typical demonstration of thermoelectricity, one side of the material is kept at one temperature and the other side at a different temperature, and the result is a *permanent* voltage across the crystal.

Mathematical Description

The pyroelectric coefficient may be described as the change in the spontaneous polarization vector with temperature:

$$p_i = \frac{\partial P_{S,i}}{\partial T}$$

where p_i ($Cm^{-2}K^{-1}$) is the vector for the pyroelectric coefficient.

Pyroelectricity can be visualized as a triangle of kinetic, electrical and thermal energy, where each corner represents energy states in the crystal. The side between electrical and thermal corners represents the pyroelectric effect and produces no kinetic energy. The side between kinetic and electrical corners represents the piezoelectric effect and produces no heat.

Although artificial pyroelectric materials have been engineered, the effect was first discovered in minerals such as tourmaline. The pyroelectric charge in minerals develops on the opposite faces of asymmetric crystals. The direction in which the propagation of the charge tends toward is usually constant throughout a pyroelectric material, but in some materials exhibiting ferroelectricity this direction can be changed by a nearby electric field. All pyroelectric materials are also piezoelectric since the two properties being closely related. But some piezoelectric materials have a crystal symmetry that does not allow pyroelectricity.

Very small changes in temperature can produce an electric potential due to a materials' pyroelectricity. Passive infrared sensors are often designed around pyroelectric materials, as the heat of a human or animal from several feet away is enough to generate a difference in charge.

The Pyroelectric Crystal Classes and Piezoelectricity

Crystal structures can be divided into 32 crystal classes according to the number of rotational axes and reflection planes they exhibit. Of the these 32 crystal classes, twenty-one are non-centrosymmetric (not having a center of symmetry), Of these, twenty exhibit direct piezoelectricity, the remaining one being the cubic class 432. Ten of these twenty piezoelectric classes are polar. They possess a spontaneous polarization, having a dipole in their unit cell and exhibit pyroelectricity. If this dipole can be reversed by the application of an electric field, the material is called as ferroelectric. When an electric field is applied, any dielectric material develops a dielectric polarization (electrostatics). However, even in the absence of a field, a polar material has such a natural charge separation. Whether or not a material is polar is determined by its crystal structure. Only 10 of the 32 point groups are polar. All polar crystals are pyroelectric, so the 10 polar crystal classes are sometimes referred to as the pyroelectric classes.

- Piezoelectric crystal classes: 1, 2, m, 222, mm2, 4, -4, 422, 4mm, -42m, 3, 32, 3m, 6, -6, 622, 6mm, -62m, 23, -43m
- Pyroelectric: 1, 2, m, mm2, 3, 3m, 4, 4mm, 6, 6mm

The pyroelectricity property can be given by measuring the change in net polarization (a vector) proportional to a change in temperature. The total pyroelectric coefficient measured at constant stress is the sum of the pyroelectric coefficients at constant strain (primary pyroelectric effect) and the piezoelectric contribution from thermal expansion (secondary pyroelectric effect). Under normal circumstances, polar materials may not display a net dipole moment. As a consequence, there are no electric dipole equivalents of bar magnets, since the intrinsic dipole moment is neutralized by "free" electric charge building up on the surface by internal conduction or from the ambient atmosphere. Polar crystals can only reveal their nature when the balance with the compensating surface charge is perturbed.

1.6. Ferroelectric Photovoltaics

The study of ferroelectric photovoltaic (PV) effects has become an important topic in the fields of ferroelectricity, condensed matter physics and materials sciences in recent years. For the rapidly growing needs in solar energy, the interest in developing ferroelectric PV devices with specific properties has drastically increased. This part first focuses on the fundamental aspects of the PV effect in ferroelectric materials, and presents a review on the latest developments in ferroelectric PV cells, to describe the primary characteristics of ferroelectric PV cells and to achieve an understanding for their future development.

Fundamentals

The global energy problem is rapidly intensifying due to escalating competition for resources from emerging, populous countries and compelling evidence pointing towards the imperative need for controlling greenhouse gas and carbon emissions. The recent nuclear-energy safety crisis in Japanese nuclear power facilities induced by an earthquake, together with the fundamental finite nature of fossil fuels, have led to urgent demands and increased interests for renewable, clean and safe energy, such as solar energy that is obtained via solar cell. This cell directly converts the sunlight energy into electric power, which has emerged as one of the most sustainable and environmentally friendly approaches for solving energy source issues [97].

The process in the solar cells for absorbing sunlight and converting it into electricity is called the photovoltaic effect (photo-for-light, and voltaic-for-voltage), and the amount of electricity power produced varies with the amount of light and its intensity, as shown in Figure **1.7.** In conventional Si semiconductor solar cells or photovoltaic (PV) cells, photons with energy higher than the band-gap are absorbed to produce electron-hole pairs, which are separated in the p–n junction [98]. The photo generated electron–hole pairs are separated by the electric field in the depletion region; then the electrons and holes move out of this region under the acceleration induced by the electric field. The open-circuit voltage (V_{oc}) in this case is limited by the band-gap (E_g) of the semiconductor, with eV_{oc} (e is the elementary electric charge) being normally less than Eg.

A variety of semiconductor types can be used to generate a photovoltaic effect, including single and polycrystalline silicon, thin-film amorphous silicon, and several advanced materials. The conventional silicon-based solar cells exhibit disadvantages of high-cost, complicated processing requirements, property instability, etc. To solve this problem, there are currently many research groups active in the field of photovoltaic cells in universities and research institutions around the world. This research can be divided into three areas: making current solar cells cheaper and/or more efficient to effectively compete with other energy sources; developing new technologies based on new solar cell architectural designs; and developing new materials to serve as light absorbers and charge carriers, among which the ferroelectrics bear the potential for generating the photovoltaic effect.

Since several decades ago, a unique photovoltaic effect had been discovered in some ferroelectrics [96]. In oxide perovskite ferroelectrics possessing spontaneous polarization, a steady photovoltaic voltage and current could exist in a homogeneous medium under uniform illumination, to produce such a ferroelectric PV effect, making ferroelectrics new materials to be used to develop solar cells. This PV effect, being fundamentally different from that based on Si, exhibits fascinating advantages such as extremely large photo-voltages (several orders of magnitude greater than the band-gap and can reach 1000-100000V), a photocurrent proportional to the polarization magnitude, and charge-carrier separation in domain walls.

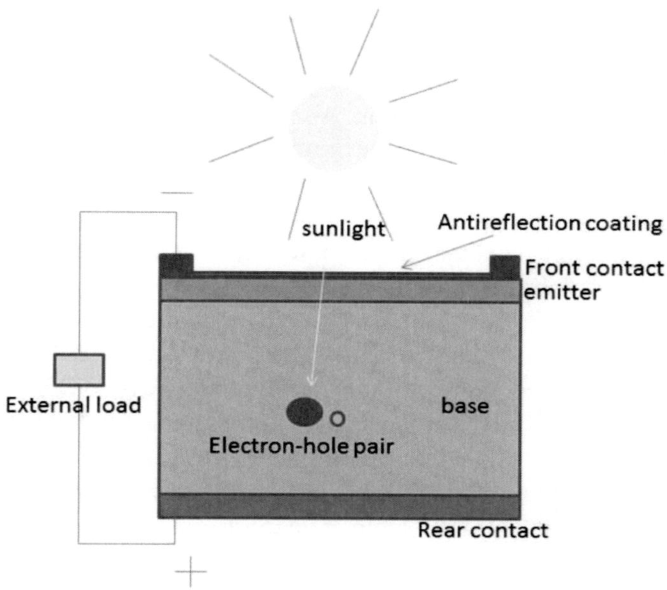

Figure 1.7. (color only website) the basic work principle of Si-based solar cells.

Materials and Methods

The definition of ferroelectric photovoltaic effect can be described as both the steady-state short circuit photovoltaic current and the anomalous open-circuit photo-voltage (PV, $V >> E_g$) mentioned above. Particularly, the observed transient photocurrents in a ferroelectric may be related to the influence of non-equilibrium photo-excited carriers (such as impurity centers) on the spontaneous polarization P_0. The corresponding voltage may give rise to the observed photorefractive effect in ferroelectrics, in which illumination may change the optical birefringence in ferroelectrics. Therefore, one possible mechanism of the ferroelectric PV effect is associated with the photorefractive effect and a change in the spontaneous polarization due to illumination, and is accompanied by a transient screening current. However, in the majority of the ferroelectrics investigated, the PV effect is due to the steady-state photovoltaic current. In spite of this, the nature of the photovoltaic effect in ferroelectrics is far from fully understood and only the first steps have been taken in developing the theory of the PV effect. A detailed and systematic account of the experimental investigations and a comparison with the proposed possible mechanisms on the assumption have been described [99].

Considering measurements for the photo-voltages in ferroelectrics under open-circuit conditions (PV effect), photo-voltages of $V >> E_g$ magnitude had been previously observed in an investigation of photoelectric phenomena in ferroelectric SbSiBr and $BaTiO_3$ single crystals and the PV effect was discovered in lithium niobate and several other ferroelectric niobates [3, 4-15]. Special attention was paid to the relationship between the PV effect and the steady-state photovoltaic current. In the case of ferroelectric niobates (such as Fe: $LiNbO_3$ and $SbNbO_3$), the photo-voltage along the spontaneous polarization (001) direction was several times greater than the band-gap. For the case of SbSiBr and $BaTiO_3$ single crystals, an initially polarized single-domain crystal was illuminated in the ferroelectric phase with light corresponding to the photosensitivity maximum in which illumination produced a short-circuit photocurrent along the (00l) direction. After the illumination, the crystal was heated to the paraelectric phase across the Curie point, and the Curie point was shifted toward higher temperatures due to the internal space-charge field. When this crystal was again illuminated in the paraelectric phase, a depolarization photocurrent was produced to make the Curie temperature shift back to the equilibrium value, from which the space-charge density and the internal field or photo-voltage could be estimated.

For the kinetics of PV effects in these crystals, a simple relationship link between the photorefractive and PV effects in the majority of the ferroelectrics has been investigated. The PV effect is inversely proportional to the photo-conductivity of ferroelectrics, and the illumination-induced change in birefringence is a consequence of the linear electro-optic effect caused by the field *E,* which appears in the PV process. The ferroelectrics may also exhibit anomalously high photo-voltages associated with the steady-state photovoltaic current depends on the crystals. [100-114] As summarized in Ref.3, there was no controversy in the early experimental works on the PV effect. A steady-state photovoltaic current generated by uniform illumination of ferroelectrics without an external field was reliable. This photovoltaic current indicated that the charging of a crystal were associated with the anomalous photo-voltages in ferroelectrics, which are several orders of magnitude greater than the band-gap. New carrier transport mechanisms allowing for the explanation of the photovoltaic current should be hypothesized based on the characteristics of the excitation, recombination, and scattering in a pyro-electric crystal. There is no doubt that studies of the PV effect, will make an important contribution to the physics of ferroelectricity, which is of intrinsic interest.

However, the ferroelectric PV effect had been neglected for a long time since most oxide ferroelectrics have a band-gap of at least 3eV, absorbing primarily in the ultraviolet (UV) region, which comprises only 8% of the solar spectrum [108]. Very recently, a big step forward in this research had been achieved in $BiFeO_3$ ferroelectrics with a smaller band-gap (2.67eV) within the visible range of 2.5-2.9 eV, in which a substantial visible-light photovoltaic effect was observed [109-116]. $BiFeO_3$ exhibits a large remnant ferroelectric polarization and a relatively high light absorption coefficient. The photo-voltage observed in $BiFeO_3$ nanometer thin films was significantly higher than its electronic band-gap [115]. Although the development of PV cells based on silicon and other functional materials has been achieved, the use of ABO_3 ferroelectric PV cells seems to provide another potential pathway for further boosting conversion efficiency and lowing cost, and may lead to a new future trend for solar-energy research [116].

Ferroelectric BiFeO3 (BFO) exhibits a rhombohedral R3c structure with a perovskite pseudo cubic unit cell ($a \approx 3.96$Å, $\alpha \approx 89.4°$) and the polarization elongate along the (111) direction below Curie temperature (Tc) ≈ 1100K, and also contains transition metal ions with unpaired *d* electrons. Their *d* electrons can result in a relatively small optical gap and give rise to a high concentration of charged impurities and defects. The electronic transport properties and a diode effect associated with the direction of bulk electric polarization had been recently discovered in $BiFeO_3$ crystals with a single ferroelectric domain [118]. For typical p-n junction diodes, the current density follows an exponential relationship of $J \propto \exp(qV/\alpha k_B T)$, where q is the electron charge, V is voltage, α is a constant called the ideality factor, and k_B is the Boltzmann constant. There exist substantial currents that are nonlinear with applied electric field E and also depend strongly on the direction of E in in BFO crystal specimens. The magnitude of E is much less than ferroelectric coercivity, thus polarization switching does not occur during the E sweep for curves of current density (J) versus E. But at 300 and 350K, BFO shows linear-scale J (E) curves, which exhibit diode like behavior, from which the diode directions switch whenever ferroelectric polarization is reversed by external pulses, and the diode forward direction is always along the ferroelectric polarization direction. The observed diode like behavior of *dc* conduction implies a possibility of zero-bias PV effect, and in BFO, a substantial PV effect with semitransparent Au electrodes illuminated with visible light also had been found, with a total power density being less than 20mW/cm^2, showing that visible light induces substantial photocurrent.

When a ferroelectric is under intense light illumination, the second-order optical response combined with a linear term induces an asymmetric polarization, which results in an optical rectification effect. This effect is an intrinsic bulk effect and becomes strongest when the polarized-light electric field is along the ferroelectric polarization and follows a sinusoidal angular dependence. On the other hand, a noticeable variation in the magnitude of the rectification and PV effects in different samples indicates the importance of impurities and defects with respect to the transport mechanism. This investigation reveals an unusual charge conduction behavior in leaky ferroelectrics (such as BFO), and opens advance studies of BFO-based multifunctional devices [118]. Although the bulk photovoltaic effect in ferroelectric BFO materials have been mainly studied in research activities, the underlying mechanism was not very clear. It was generally believed that the effect was associated with the loss of inversion symmetry in the distribution of defects, impurities, space charges and polarizations in ferroelectrics.

Recently, a very high photovoltaic voltage originated from the domain walls of the ferroelectric material was observed in BFO thin films by Yang *et al.* BFO films with thicknesses between 100 and 500nm were grown on single-crystalline (110) $DyScO_3$ (DSO) substrates by metal–organic chemical vapor deposition (MOCVD). Through careful design of the experiments by using $BiFeO_3$ thin films of varying thicknesses and domain types, they convinced unambiguously that the photovoltaic effect in $BiFeO_3$ was originated from the 71°and/or 109° domain walls, owing to the asymmetry in polarization, instead of bulk photovoltaic effect [116].

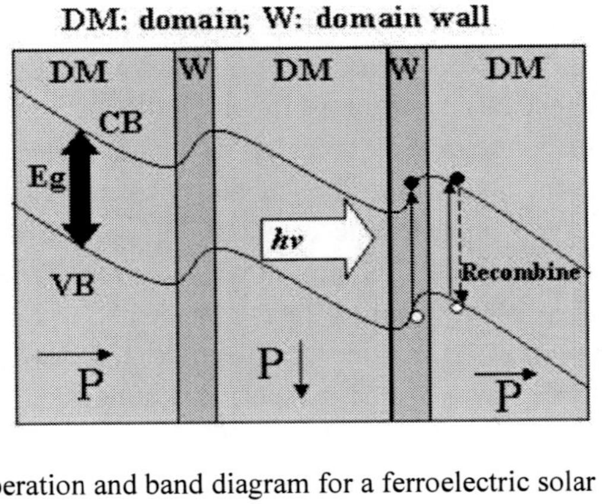

Figure 1.8. Principles of operation and band diagram for a ferroelectric solar cell, the band bending occurs across the domain walls.

The most evident difference between the photovoltaic effect in a ferroelectric material and that in a conventional semiconductor p-n junction is the magnitude of the electric field that separates the photo-generated electron-hole pair. The effective electric field in a ferroelectric material is around one order of magnitude higher than that in a p-n junction. The domain wall in BFO serves the function as a p-n junction, as shown in Figure 1.8. [118] Since the polarization divergence at the domain walls is non-zero (that is, $\mathbf{P} \neq 0$) and the imbalance charges accumulates at the domain walls, inducing an electrostatic potential offset varying from 0.02eV to 0.20eV in various materials, which causes a band bending across the domain wall, similar to the band bending in the depletion layer of a typical p-n junction. Therefore, the stronger electric field and efficient charge separation occurs at the domain wall (as thin as 2nm) in films. While in a domain in bulk $BiFeO_3$, the photo-induced charge carriers could be localized and tightly bound, allowing them to quickly recombine before their diffusing to either side of a domain wall. This may explain why the bulk photovoltaic effect in $BiFeO_3$ is negligibly small. On the other hand, the domain wall photovoltaic effect can be deduced from the fact that the open-circuit voltage increases linearly with the number of domain walls, which implies that a high voltage output in $BiFeO_3$ can be obtained from well-arranged multiple domain walls. In contrast, a single-domain sample cannot generate appreciable change in open-circuit voltage with increasing sample thickness.

But unfortunately, the power conversion efficiency of an ferroelectric solar cell based on BFO thin film is still poor (about a degree of 0.125% [117]) and much lower than that of commercially available solar cells. The main reason is because the leaky ferroelectric domains in BFO cannot withstand strong electric polarization and it is quite difficult to increase the bulk conductivity of ferroelectric domains, and also difficult to achieve high conversion efficiency. Furthermore, the polarization-associated charge density across a domain wall in leaky ferroelectrics will be reduced for lowering the open-circuit voltage.

Although, $BiFeO_3$ (BFO), as a multi-ferroic material, [101-104] in which the ferroelectric, ferromagnetic and ferro-elastic orderings interact with each other, it can provide a novel opportunity to tune the output signal of a solar cell by various mechanical, electrical and magnetic means. Yao et al. have already carried out some preliminary attractive experiments in applying an electric field higher than the coercive field in the manipulation of ferroelectric domain structures by an external field, to turn on and off the photovoltaic voltage [103]. They prepared BFO thin films by RF magnetron sputtering and deposited the films on STO (001) substrates, in which appreciable photovoltaic response in the visible waves could be observed. The as-deposited BFO films are self-polarized and produce large open-circuit photo voltage up to 0.3V. The large portion of the photo-voltage and photocurrent (about two thirds) is switchable in response to the switching of the ferroelectric polarization, with the direction of the photocurrent opposite to that of the polarization vector, which is the main driving force of the observed photovoltaic effect. Additionally, an important factor which determines the depolarization field—screening effect in ferroelectrics could be considered in influencing the ferroelectric photovoltaic effect. Qin et al [118] recently had studied the mechanism of the photovoltaic effect in ferroelectrics by considering both ferroelectric materials and their electrodes as a whole system, and suggested that the oxide electrodes especially with high dielectric constant could provide a larger photovoltaic response.

Due to their low cost and simple fabrication process, polycrystalline BFO (poly-BFO) films also show their potential for practical applications rather than other BFO epitaxial films. However, the photovoltaic efficiency of the polycrystalline films may be lower than that of epitaxial films. Thus, it is necessary to study and improve the photovoltaic effect in polycrystalline ferroelectric films. The study of Poly-BFO thin films on Pt/Ti/SiO_2/Si substrates by the sol–gel method was reported in Ref. [117], which indicated that the oxide electrodes provide a larger photovoltaic response, and a Sn doped In_2O_3 (ITO)/poly-BFO/Pt capacitor has almost 25 times larger photovoltaic response than that of an Au/Poly-BFO/Pt capacitor. Moreover, an almost 1000-fold photo-conductivity enhancement is obtained in an ITO/poly-BFO/Pt capacitor. These experimental results reveal that a combined system of narrow-band-gap ferroelectric BFO and oxide ITO electrodes can give a considerably high photovoltaic response. Furthermore and very recently, to efficiently separate the electrons and holes and thus prevent their recombination in organic photovoltaic devices, a ferroelectric polymer layer with a large, permanent, internal electric field was reported to be incorporated into the device, which enhanced the efficiency of several types of organic photovoltaic device from 1–2% without layers to 4–5% with layers. [119]

REFERENCES

[1] Jacques et Pierre Curie, *Compt. rend.* 91, 294-295, 383-387 (1880).

[2] A.M. Nicholson, *Piezophony. U.S. Patent* 1, 495, 429. Filed April 10, 1918. Patented May 27, 1924. A.M. Nicholson, Generating and Transmitting electric Currents., U.S. Patent 2, 212, 845. Filed April 10, 1918. Patented August 27, 1940.

[3] Paul Langevin, Procédé et appareils d'émission et de réception des ondes élastiques sous-marines à l'aide des propriétés piézoélectriques du quartz. Brevet d'Invention No. 505, 703. Demandé le 17 septembre 1918. Délivré le 14 mai 1920.

[4] P. Debye, Physik. Zeitschr. XIII, 97-100 (1912). Einige Resultate einer kinetischen Theorie der Isolatoren (Vorläufige Mitteilung).

[5] Erwin Schrödinger, Aus sen Sitzungsberichten der Kaiserl. Akademie der Wissenschaften in Wien. Mathem.-Naturw. Klasse; Bd. CXXI, Abt. Iia, November 1912. Studien üder Kinetik der Dielektrika, dam Schmelzpunkt, Pyro-und Piezoelekrizität.

[6] J. Valasek, *Phys. Rev.* 15, 537 (1920); 17, 475-481 (1920). Piezoelectric and allied phenomena in Rochelle salt.

[7] J. Valasek, *Phys. Rev.* 19, 478-491 (1922); Piezo-electric activity of Rochelle salt under various conditions.

[8] G. Steulmann, *Z. Phys.* 77, 114-116 (1932). Die Dielektrizitätskonstante einer Anzahl von Kaliumsalzen und Alkalihalogeniden.

[9] J. West, *Z. Krist.* 74, 306 (1930). A quantitative X-Ray Analysis of the Structure of Potassium-Dihydrogen-Phasphate (KH2PO4).

[10] G. Busch and P. Scherrer, *Naturwiss.* 23, 737 (1935). Eine neue Seignetteelektrische Substanz.

[11] G. Busch, *Helv. Phys. Acta* 11, 269-298 (1938). Neue Seignette-Elektrika.

[12] J.C. Slater, *J. Chem. Phys.* 9, 16-33 (1941). Theory of the Transition of KH2PO4.

[13] H. Thurnaurer and J. Deaderick, *U.S. Patent* 2, 429, 588, Oct. 21, 1947, filed Oct. 2 1941.

[14] A. von Hippel, R.G. Breckenbridge, F.G. Chesley, and L. Tisza, *Ind. Eng. Chem.* 38, 1097 (1946).
[15] B. Wul and J.M. Goldman, *C.R. Acad. Sci. URSS* 51, 21 (1946).
[16] W. Känzig, *Helv. Phys. Acta* 24, 175 (1951).
[17] A.F. Devonshire, *Phil. Mag.* 40, 1040 (1949), *Phil. Mag.* 42, 1065 (1951).
[18] H.D. Megaw, *Ferroelectricity in Crystals*. London: Methuen and Co., Ltd., 1957, p. 2.
[19] J. F. Nye, *Physical Properties of Cyrstals,* Oxford University Press, Oxford, (1985).
[20] R. B. Gray, "Transducer and Method of Making the Same," *USA Patent* No. 2, 486, 560 (Nov 1, 1949 1949).
[21] S. Roberts, "Dielectric and Piezoelectric Properties of Barium Titanate, " *Physical Review*, 71 [12] 890-5 (1947).
[22] L. A. Shuvalov, "Symmetry Aspects of Ferroelectricity," *Journal of the Physical Society of Japan*, 28, 38-51 (1970).
[23] Richard E. Eitel, Novel Piezoelectric Ceramics: Development Of High Temperature, High Performance Piezoelectrics On The Basis Of Structure, thesis, 2003, The Pennsylvania State University.
[24] Brian joseph rodriguez, nanoscale investigation of the piezoelectric properties of perovskite ferroelectrics and III-nitrides, thesis, 2003, North Carolina State University.
[25] M. E. Lines and A. M. Glass, Principles and Applications of Ferroelectrics and Related Materials (Oxford Science Publications, 1977).
[26] W. Merz, *Phys. Rev.* 95, 690 (1954).
[27] J.E. Ralph, J.P. Gowers and M.R. Burgess, *Appl. Phys. Lett.* 41, 343-345 (1982).
[28] G. Rosenman, D. Shur, *J. Appl. Phys.* 88, 6109-6161 (2000).
[29] S. Miyazawa, *J. Appl. Phys.* 50, 4599-4603 (1979).
[30] F. Saurenbach and B. D. Terris, *Appl. Phys. Lett.* 56, 1703-1705 (1990).
[31] H. Birk, J. Glatz-Reichenbach, L. Jie, E. Schreck, and K. Dransfeld, *J. Vac. Sci. Technol.* B 9, 1162-1165 (1991).
[32] R. Luthi, H. Haefke, K.-P. Meyer, E. Meyer, L. Howald, and H.-J. Guntherodt, *J. Appl. Phys.* 74, 7461-7471 (1993).
[33] M.-K. Bae, T. Horiuchi, K. Hara, Y. Ishibashi and K. Matsushige, *Jpn. J. Appl. Phys.* 33, 1390-1396 (1994).
[34] L. Eng, M. Friedrich, J. Fousek and P. Gunter, *J. Vac. Sci. Technol.* B 14, 1191-1196 (1996).
[35] S. V. Kalinin, C. Y. Johnson, and D. A. Bonnell, *J. Appl Phys*. 91, 3816–3823 (2002).
[36] S. V. Kalinin and D. A. Bonnell, *Appl. Phys. Lett.* 78, 1116-1118 (2001).
[37] P. Guthner, and K. Dransfeld, *Appl. Phys. Lett.* 61, 1137-1139 (1992).39.
[38] A. Gruverman, O. Auciello and H. Tokumoto, *J. Vac. Sci. Technol.* B 14, 602-605 (1996).
[39] T. Hidaka, T. Maruyama, I. Sakai, M. Saitoh, L.A. Wills, R. Hiskes, S.A. Dicarolis and J. Amano, *Integrated Ferroelectrics* 17, 319-327 (1997).
[40] C.S. Ganpule, V. Nagarajan, B.K. Hill, A.L. Roytburd, E.D. Williams, R. Ramesh, S.P.Alpay, A. Roelofs, R. Waser, and L.M. Eng, *J. Appl. Phys.* 91, 1477-1481 (2002).
[41] G. Zavala, J.H. Fendler and S. Troiler-McKinstry, *J. Appl. Phys.* 81, 7480-7491 (1997).
[42] K. Franke, J. Besold, W. Haessle, C. Seegebarth, *Surf. Sci. Lett.* 302, L283-L288 (1994).

[43] C.S. Ganpule, A.L. Roytburd, V. Nagarajan, B.K. Hill, S.B. Ogale, E.D. Williams, R. Ramesh, and J.F. Scott, *Phys. Rev.* B 65, 014101-108 (2002).

[44] A. Kholkin, *Ferroelectrics* 221, 219-228 (1999).

[45] A. K. Tagantsev, I. Stolichnov, E. L. Colla, and N. Setter, *J. Appl. Phys.* 90, 1387 (2001).

[46] C. H. Park and D. J. Chadi, *Phys. Rev.* B 57, R13961 (1998).

[47] W. Y. Pan, C. F. Yue, K. W. Lin, S. Sun, and B. A. Tuttle, *Journal of Materials Science Letters* 12, 986 (1993).

[48] M.-J. Pan, S.-E. Park, C. W. Park, K. A. Markowski, S. Yoshikawa, and C. A. Randall, *J. Am. Ceram. Soc.* 1996, 2971 (1996).

[49] C. Brennan, Ferroelectrics 150, 199 (1993).

[50] W. L. Warren, D. Dimos, B. A. Tuttle, R. D. Nasby, and G. E. Pike, *Appl. Phys. Lett.* 65, 1018, (1994).

[51] H. N. Al-Shareef, D. Dimos, T. J. Boyle, W. L. Warren, and B. A. Tuttle, *Appl. Phys. Lett.* 68, 690 (1996).

[52] D. Dimos, W. L. Warren, M. B. Sinclair, B. A. Tuttle, and R. W. Schwartz, *J. Appl. Phys.* 76, 4305 (1994).

[53] S. D. Bernstein, T. Y. Wong, Y. Kisler, and R. W. Tustison, *J. Mater. Res.* 8, 12 (1992).

[54] T. Mihara, H. Watanabe, and C. A. P. d. Araujo, *Jpn. J. Appl. Phys.* 33, 3996 (1994).

[55] S. B. Desu, *Phys. Stat. Sol.* (a) 151, 467 (1995).

[56] S. Aggarwal and R. Ramesh, *Annu. Rev. Mater. Sci* 28, 463 (1998).

[57] D. P. Vijay and S. B. Desu, *Journal of the Electrochemical Society* 140, 2640 (1993).

[58] J. Lee, L. Johnson, A. Safari, R. Ramesh, T. Sands, H. Gilchrist, and V. G. Keramidas, *Appl. Phys. Lett.* 63, 27 (1993).

[59] H. N. Al-Shareef, K. R. Bellur, A. I. Kingon, and O. Auciello, *Appl. Phys. Lett.* 66, 239 (1995).

[60] E. L. Colla, D. V. Taylor, A. K. Tagantsev, and N. Setter, *Appl. Phys. Lett.* 72, 2478 (1998).

[61] E. L. Colla, A. K. Tagantsev, D. V. Taylor, and A. L. Kholkin, *Integr. Ferroelectr.* 18, 19 (1997).

[62] E. L. Colla, S. Hong, D. V. Taylor, A. K. Tagantsev, N. Setter, and K. No, *Appl. Phys. Lett.* 72, 2763 (1998).

[63] E. L. Colla, A. K. Tagantsev, A. L. Kholkin, and N. Setter, *Integr. Ferroelectr.* 10, 289 (1995).

[64] X. Du and I.-W. Chen, *J. Appl. Phys.* 83, 7789 (1998).

[65] I. K. Yoo and S. B. Desu, *Materials Science and Engineering* B 13, 319 (1992).

[66] Kwangbae Lee and Byung Roh Rhee, Leakage Current-Voltage Characteristics of Ferroelectric Thin Film Capacitors, *Journal of the Korean Physical Society*, Vol. 38, No. 6, June 2001, pp. 723728.

[67] R. Moazzami, C. Hu and W. Shepherd, in *Proceedings International Reliability Physics Symposium*, (1990), p. 231.

[68] C. Sudhama, A. C. Campbell, P. D. Manier, R. E. Jones, R. Moazzami, C. J. Mogab and J. C. Lee, *J. Appl. Phys.* 75, 1014 (1994).

[69] J. F. Scott, B. M. Melnick, J. D. Cuchiaro, R. Juleeg, C. A. Paz de Araujo, A. D. McMillan and M. C. Scott, *Intergr. Ferroelectr.* 4, 85 (1994).

[70] T. Mihara and H. Watanabe, *Jpn. J. Appl. Phys.* 34, 5664 (1995).

[71] I. Stolichnov and A. Tagantsev, *J. Appl. Phys*. 84, 3216, (1998).
[72] *Britannica*. Dielectrics (Physics) 2009. 2009-08-12.
[73] Arthur R. V. Hippel, Technology Press of MIT and John Wiley, NY, 1954.
[74] John T. S. Irvine, Derek C. Sinclair, and Anthony R. West, Electroceramics: Characterization by Impedance Spectroscopy, *Adv. Mater*., *2* (1990) No. 3, 132.
[75] C. Z. Rosen, B. V. Hiremath, and R. Newnham, *Piezoelectricity* (American Institute of Physics, 1992).
[76] K.-H. Hellwege and A. M. Hellwege, Landolt-Börnstein Tables, edited by K.-H. Hellwege (Springer-Verlag, Berlin, 1969), Vol. 11.
[77] W. P. Mason and H. Jaffe, *Proceedings of the IRE*, 921 (1954).
[78] D.-G. Kim, ISIF, Colorado Springs, CO (1999).
[79] D. Damjanovic, *J. Appl. Phys*. 82, 1788 (1997).
[80] K. Lefki and G. Dormans, *J. Appl. Phys*. 76, 1764 (1994).
[81] H. Birk, J. Glatz-Reichenbach, Li-Jie, E. Schreck, and K. Dransfield, *J. Vac. Sci. Technol*. B 9, 1162 (1991).
[82] Lie-Jie, C. Baur, B. Koslowski, and K. Dransfield, *Physica* B 204, 318 (1995).
[83] J. A. Christman, R. R. Woolcott, A. I. Kingon, and R. J. Nemanich, *Appl. Phys. Lett*. 73, 3851 (1998).
[84] A. L. Kholkin, C. Wuetchrich, D. V. Taylor, and N. Setter, *Rev. Sci. Instrum*. 67, 1935 (1996).
[85] J.-F. Li, P. Moses, and D. Viehland, *Rev. Sci. Instrum*. 66, 215 (1995).
[86] Q. Zhang, W. Pan, and L. E. Cross, *J. Appl. Phys*. 63, 2492 (1988).
[87] Q. Zhang, S. Jang, and L. E. Cross, *J. Appl. Phys*. 65, 2807 (1989).
[88] D. Royer and V. Kmetik, *Electronics Letters* 28, 1828 (1992).
[89] G. Arlt, H. Dederichs and R. Herbiet, "90o-Domain Wall Relaxation in Tetragonally Distorted Ferroelectric Ceramics," *Ferroelectrics*, 74 37-53 (1987).
[90] K. Uchino and S. Hirose, "Loss Mechanisms in Piezoelectrics: How to Measure Different Losses Separately," *IEEE Transactions on Ultrasonics, Ferroelectrics, and Frequency Control*, 48 [1] 307-21 (2001).
[91] P. Gerthsen, K. H. Hardtl and N. A. Schmidt, "Correlation of Mechanical and Electrical Losses in Ferroelectric Ceramics," *Journal of Applied Physics*, 51 [2] 1131-4 (1980).
[92] Webster, John G (1999). *The measurement*. pp. 32–113. ISBN 978-0-8493-8347-2.
[93] Earle R. Caley and John F.C. Richards, *Theophrastus: On Stones* (Columbus, Ohio: Ohio State University, 1956), page 51.
[94] Johann Georg Schmidt, *Curiöse Speculationes bey Schalflosen Nächten* [Curious Speculations During Sleepless Nights] (Chemnitz and Leipzig (Germany): Conrad Stössen, 1707), pages 269-270.
[95] Brewster, David (1824). "Observations of the pyro-electricity. *The Edinburgh Journal of Science* 1: 208–215.
[96] Damjanovic, Dragan, 1998, Ferroelectric, dielectric and piezoelectric properties of ferroelectric thin films and ceramics, *Rep. Prog. Phys*. 61, 1267–1324.
[97] Ginley, D., Green, M. A., Collins, R. (2008) Solar energy conversion toward 1 terawatt. *Mater. Res. Soc. Bull*. 2008; 33: 355–364.
[98] Gur, I., Fromer, N. A., Geier, M. L., Alivisatos, A. P. (2005) Air-stable all-inorganic nanocrystal solar cells processed from solution. *Science* 2005; 310: 462–465.

[99] Fridkin, V. M., Popov, B. N. (1978) Anomalous photovoltaic effect in ferroelectrics. *Usp. Fiz. Nauk* 1978;126: 657-671.

[100] Joseph, W. B., Ilya G., Andrew M. R. (2008) New Highly Polar Semiconductor Ferroelectrics through *d* Cation-O Vacancy Substitution into PbTiO3: A Theoretical Study. *J. Am. Chem. Soc.* 2008; 130: 17409–17412.

[101] Yang, S. Y., Seidel, J., Byrnes, S. J., Shafer, P.C., Yang, H., Rossell, M. D., Yu, P., Chu, Y.-H., Scott, J. F., Ager, J.W., Martin, L. W., Ramesh, R. (2010) Above-bandgap voltages from ferroelectric photovoltaic devices. *Nature Nanotechnology* 2010; 5: 143-147.

[102] Huang H. (2010) Ferroelectric photovoltaics. *Nature Photonics* 2010; 4:143-145.

[103] Ji, W., Yao, K., Liang, Y. C. (2010) Bulk Photovoltaic Effect at Visible Wavelength in Epitaxial Ferroelectric BiFeO3 Thin Films. *Advanced materials* 2010; 22: 1763-1766.

[104] Lee, T. C S., Choi, Y. J., Kiryukhin, V., Cheong S.-W. (2009) Switchable Ferroelectric Diode and Photovoltaic Effect in BiFeO3. *Science* 2009; 324: 63-66.

[105] Grekov A. A., Malitskaya M. A., Spitsyna V. D., and Friedkin V. M., *Kristallografiya* 1970; 15: 500.

[106] Volk T. R., Grekov A. A., Kosonogov N. A., and Malitskaya M., *Sov. Phys. Solid State* 1972;14: 3214.

[107] Glass A. M., Von der Linde D., Auston D. H., and Negran T. J. *J. Electron. Mater.* 1975; 4: 915.

[108] Levanyuk A. P. and Osipoy V. V., *Izv. Akad. Nauk SSSR Ser. Fiz.* 1977; 41: 752.

[109] Fridkin M., *Ferroelectric Semiconductors*, Fizmatgiz, M., 1976.

[110] Ionov V., Verkhovskaya K. A., Ivleva L. I., and Fridkin V. M., *Kratk. Soobshch. Fiz.* 1973; 10: 24.

[111] Ionov P. V., *Sov. Phys. Solid State* 1973; 15: 2827.

[112] Glass A. M. and Von der Linde D., *Ferroelectrics* 1976; 10: 163.

[113] Giinter P. and Micheron F., *Ferroelectrics* 1978; 18: 27.

[114] Micheron. F, *Ferroelectrics* 1978; 18: 153.

[115] Barkan B., Marennikov S. I., Pestryakov E. V., and Entin M. V. *lzv. Akad. Nauk SSSR Ser. Fiz.* 1977; 41:748.

[116] Glass M., Von der Linde D., and Negran T. J., *Appl. Lett.* 1974; 25: 233.

[117] Fridkin V. M., Popov B.N., and Ionov P. V., *Ferroelectrics* 1978; 18: 165.

[118] Ionov P. V., *Izv. Akad. Nauk SSSR Ser. Fiz.* 1977; 42: 771.

[119] Popov B.N., Fridkin V. M., *Sov. Phys. Solid State* 1978; 20: 413.

Chapter 2

RELAXOR FERROELECTRICS AND FUNDAMENTALS FOR HIGH TC FERROELECTRICS

2.1. RELAXOR FERROELECTRICS

2.1.1. Definition and Properties [1]

Generally, ferroelectrics contain one or more polar axes along which a spontaneous polarization can be exhibited below the Curie temperature. The polarization direction can be forced to be changed by the application of a strong electric field. The arrangement of cations and anions within a ferroelectric gives rise to dipole moments within each unit cell, and via the material's surface current, the resulting polarization can be measured.

One distinctive characteristic of ferroelectrics is the hysteretic behavior in the polarization vs. electric field curve. A net spontaneous polarization exists even after the electric field is decreased to zero. Above the Curie temperature, the materials exhibit a paraelectric phase, where the material has higher symmetry and behaves like a normal dielectric with no hysteresis. Some ferroelectrics have additional lower symmetry ferroelectric-ferroelectric transitions below the Curie temperature, and some even undergo a transition back to a paraelectric state. Rochelle salt (sodium potassium tartrate tetrahydrate) exhibits upper and lower Curie temperatures and was the material in which ferroelectricity was first observed [2].

Eleven of the 32 possible crystal classes are centro-symmetric and hence have no unique polar axis and cannot exhibit ferroelectricity. The remaining 21 crystal classes exhibit the pyroelectric effect, which is a change in polarization when the temperature is varied, and 20 of these are piezoelectric. The direct piezoelectric effect is the development of strain in a crystal when an electric field is applied. The inverse effect, which is the development of polarization when a strain is applied, also occurs. The phenomenon of electrostriction is a second order piezoelectric effect in which the developed strain is proportional to the square of the applied electric field. Piezoelectric and electrostrictive tensors describe a material's behavior under an electric field in many important applications.

Typically, the piezoelectric effect is much larger, but in certain materials the electrostrictive effect dominates and is preferred because the strain direction is independent of field direction and the hysteresis effects are reduced. Applications range from simple acoustical devices such as microphones to advanced positioning devices for scanning microscopes. One of the most commonly used electrostrictive materials is lead magnesium niobate with the chemical formula $Pb(Mg_{1/3}Nb_{2/3})O_3$, or PMN, and solid solutions of PMN with lead titanate, $PbTiO_3$.

The ability of ferroelectrics to change their atomic structure under the influence of electric fields gives rise to extremely high values of permittivity, and hence dielectric constant. For instance, lead magnesium niobate in single crystal form has a dielectric constant just over 30,000 with the actual maximum depending on measurement frequency and temperature. Since the discovery of barium titanate and related ferroelectrics in the late 1940's, intense ferroelectric research began. These materials were widely developed as thin layer capacitors for use in integrated circuits for charge storage, bypass and coupling functions. The piezoelectric effect is the most common useful property of ferroelectrics. Lead zirconate titanate, Pb(ZrxTi1-x)O_3 or PZT, dominates this field today, but new applications and materials are always being developed.

There are two significant applications involving incorporating ferroelectrics into silicon devices: the high dielectric constant materials are being developed for increased charge storage in a smaller area in memory chips. The ability of ferroelectrics to exist in two different polarizations in the absence of a field is being used for non-volatile ferroelectric memories.

Remanent polarization is another characteristic feature of ferroelectrics. Application for a variable electric field to a ferroelectric leads to the hysteretic behavior. At high fields, a saturation value of polarization is reached. When the electric field returns to zero the system is left with a spontaneous polarization, which can be flipped by applying a large enough field in the opposite direction. In contrast, a normal dielectric has no spontaneous polarization and its P-E curve is simply a lineal line with a slope of $\chi_e\varepsilon_0$ where χ_e is the electric susceptibility or one less than the relative dielectric constant.

2.1.2. Perovskite Structure

The most commonly studied ferroelectrics have the cubic perovskite structure in the paraelectric state with chemical formula ABO_3, in ehixh A-site cations occupy the corners of a cube, while B-site cations sit in the body center. Three oxygen atoms per unit cell rest on the faces. The lattice constant of the perovskites is always close to 4 Å due to the rigidity of the oxygen octahedra network and the well-defined oxygen ionic radius of 1.35 Å. All materials in this study have structures based on the perovskite structure, with displacements resulting in polarization and leading to crystal symmetry changes. Another common ferroelectric structure is the layered perovskite structure which consists of a varying number of perovskite unit cells separated by an oxide layer. This structure is also found in many high T_C superconductors.

A practical advantage of the perovskite structure is that many different cations can be substituted on both the A and B-sites without drastically changing the overall structure. Complete solid solutions are easily formed between many cations, often across the entire range of composition. Even though two cations are compatible in solution, their behavior can be radically different when apart from each other.

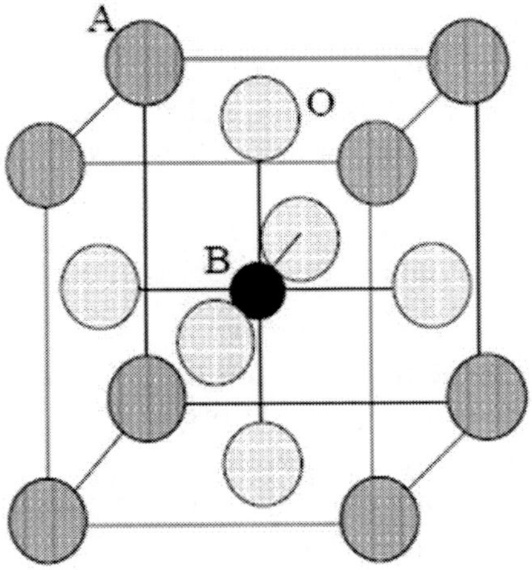

Figure 2.1. ABO_3 structure.

Thus, it is possible to manipulate a material's properties such as Curie temperature or piezoelectric constant with only a small substitution of cation. It is also possible to combine materials that would otherwise have different symmetries which can have the effect of radically altering the piezoelectic behavior. Within the perovskite system, this compatibility can be used to tailor a material's lattice constant for matching with a substrate in thin film applications.

2.1.3. Normal and Relaxor Ferroelectrics

Barium titanate ($BaTiO_3$ or BT), is the most commonly studied "normal" ferroelectric. Normal ferroelectrics are characterized by a sharp phase transition from the paraelectric to ferroelectric state in the relation of permittivity v.s temperature. In BT, the ferroelectric transition occurs at 120° C, which roughly corresponds to the peak in permittivity. Below the Curie temperature, the cubic structure changes to lower symmetry and there is a sharp decrease in dielectric constant. Above the Curie temperature (T_C), the relative permittivity follows the Curie-Weiss law.

In contrast, relaxor ferroelectrics often contain more than 2 cations and have the more gradual permittivity peaks. For instance, Lead magnesium niobate, $Pb(Mg_{1/3}Nb_{2/3})O_3$, is the prototypical relaxor ferroelectric and consists of a mixed B-site: Pb occupies the corner Asite and the B-site can be either Mg or Nb. Its very large permittivity maximum exhibits a broad transition band and frequency-dependent measurement, which are the characteristics of relaxors. By increasing the measurement frequency, the maximum of the permittivity decreases but the temperature of that maximum increases. In normal ferroelectrics, the sharp transition in permittivity also indicates a phase transition as the symmetry of the crystal changes. For relaxors, however, the broad peak in ε does not always correspond to a structural phase transition, so the temperature of the maximum permittivity is called T_M instead of T_C. In the case of PMN, there exists local polar regions well above T_M, and below T_M, polar domains continues flipping.

However, no transition to a ferroelectric state occurs even down to 4 K. Small micro domains on the order of 100 Å have rhombohedral symmetry, but the average structure remains cubic. [3, 4] By combining PMN with lead titanate, PbTiO3, solid solutions may be formed. Since lead titanate is a normal ferroelectric with large polar domains and an abrupt transition from the paraelectric to ferroelectric phase at 490° C, combining PMN with small amounts of PT can cause PMN to behave more like a normal ferroelectric; and with more than 30% concentration of PT, PMN-PT solution becomes a normal ferroelectric.

Some applications of relaxors are wide-ranging. The more gradual dependence of the permittivity on temperature and the frequency dispersion of relaxor ferroelectrics encouraged them in many applications. Both the high dielectric constant and impressive electrostrictive coefficients have been exploited. The high dielectric constants are used in ever smaller multi-layer ceramic capacitors. Relaxors are especially useful not only for their very high dielectric constants, but for those high values occur near room temperature and have a gradual temperature dependence. The relaxor ferroelectric crystals such as PMN-PT crystals, have recently been found to show excellent piezoelectric properties. Piezoelectric effects are exploited in a large number of applications, and the electrostrictive effect in materials like PMN allows for very precise control. Thus, they are employed in such applications as micropositioners, biomedical instruments and surface deformable mirrors [5].

2.1.4. Relaxor Theories [1]

Normal ferroelectrics is a convenient system to study the general properties of phase transitions, which undergo a sharp phase transition at the Curie temperature. Above T_C, each atom within the unit cell is located on a high symmetry position with no net dipole moment, and the atoms are free to vibrate about these equilibrium positions, exciting the various acoustic and optical phonon modes. The frequency of some of these modes is temperature dependent. At the phase transitions and as the temperature is lowered, the force constant of a given mode approaches zero. This is known as 'softening' of the mode and allows the atoms to shift without hindrance along that mode's direction. Spontaneous polarizations can result in a change of symmetry, since it allows for any symmetrically equivalent direction to be chosen and encourages the cooperative effects to align the other unit cells in that direction. Under an external electric field, the field direction determines the polarization direction. Domain boundaries can exist in which different grains have different dipole orientations.

But the situation is more complicated in relaxor ferroelectrics. Since there is not a uniform atomic shift throughout the entire crystal, the net polarization remains zero, while the microscopic polar domains still exist. Ye et al. had summarized that various theoretical models had been proposed to explain relaxor behavior. Smolenskially proposed that the relaxor behavior in PMN was due to varying local Curie temperatures. Since the randomness exists at the B-site, each unit cell is subjected to a different net field. Thus, there are a range of possible Curie temperatures, resulting in the broadened transition of permittivity versus temperature. This simple model has been widely cited, but now there are many other efforts to more formally explain and describe the relaxor behavior. [6-8]

Cross proposed a super paraelectric model for relaxors, which was similar to the super paramagnetic state. This model involves polar micro regions that are dynamically disordered above T_m, i.e., the mean Curie temperature of the different regions. Activated by thermal energy, the cations continuously flip between equivalent directions. The local symmetry is lower than the global, since the heterogeneity caused by the mixed B-site creates locally favorable directions. However, the small energy barrier separating the different directions resulted in that the macroscopic polar domains never form the normal ferroelectrics. [9]

In another possible model for dipolar glass-like behavior, Viehland et al [10] analyzed the dielectric response using the Vogel-Fulcher relationship [11] which also implies activated dynamics and showed the evidence for glassy behavior. As the evidence seen in the very slow dynamics, the glassy nature could be believed to be due to the randomly oriented dipolar fields. Differently from the super paraelectric model, the cations are frozen along a particular <111> direction. Another theory related to the contributions from a quenched random field is also due to the compositional heterogeneity. It was proposed that PMN's broad phase transition was due to random field interactions, also causing freezing into nanometric domains. The charge heterogeneity within relaxors such as PMN and the relation to the ferroelectric domains is helpful to understand the behavior of this relaxor.

In general, the B-site chemical heterogeneity can be believed to play an important role in the diffuse phase transition in relaxor ferroelectrics. By adding dopants to the A and B-sites, the chemical order on the B-site can be changed. This can provide a measure and knowledge of how the order affects the relaxor behavior and structure.

2.1.5. Thin Films [1]

The ferroelectric ceramic is the most common form of ferroelectrics for both research and applications. Ceramics have the advantage that they are easy to form and can be pressed into whatever shape is necessary. In comparison, single crystals are more difficult to grow and are generally too expensive for applications, although they are used in research as their material properties tend to be better and their orientational effects can be studied. In the past decades, there has been interest in ferroelectric thin films in both research and application. Thin films can exhibit material properties that reach to those of the bulk in the limit of thick samples. However, they have possibly higher breakdown voltages. Other advantages of the thin films include the ability to make very small devices which can be densely packed. With ferroelectric elements incorporated with silicon-based integrated circuits, "smart" devices can be developed, which would use the ferroelectrics to act as sensors, and send information to a feedback unit. The device could then respond to its environment using the actuation properties. Thin films provide a means of studying both surface and interfacial effects. These effects may affect a material's properties and phase transition behavior.

Many advanced film deposition techniques contribute to drive the growth in thin films. The vapor phase deposition and liquid phase chemical processing techniques are the two general classes for depositing ferroelectrics onto substrates. The former allows for careful control of film thickness, orientation and stoichiometry but is often more costly and time consuming. The later techniques are low cost and faster, and often produce films with better properties, although the crystalline quality of the film is inferior.

Typically, thin film growth occurs at lower temperatures than single crystal or ceramic sintering temperatures. This is important for compatibility with a given substrate. For instance, Pb(Zrx, Ti1-x)O3 (PZT) is the most commonly used material for piezoelectric applications in both bulk and thin film form. Complete solid solutions of the parent compounds, lead zirconate and lead titanate, are formed across the entire composition range. The best piezoelectric properties are found near the morphotropic phase boundary at $x = 0.47$. In this area, there is a chemical compositional transition from rhombohedral to tetragonal phases, and materials on this "morphotropic phase boundary" exhibit the highest piezoelectric coefficients.

2.1.6. Phase Transitions for Relaxor Ferroelectrics

In recent years, relaxor ferroelectrics were discovered to be materials with large dielectric susceptibility, and have found applications in industry with optical applications such as phase conjugate mirrors, piezoelectric sensors and actuators [12]. The understanding of the ferroelectric nature of the materials arises from the competition between order-disorder due to the structural composition, and this in turn can lead to dramatic static and dynamical properties in the material [13], although the nature of the transition and composition of the relaxor state are debated. The relaxor ferroelectrics can be categorized into two structural families: perovskites with complex composition and bronze tungsten structure [14]. This field of research has many open questions. Due to the chemical disorder and lattice defects there exist dipoles which can polarize the region around them forming nano/microdomains [13]. In the relaxor there is a distinct transition from high temperature with polar nanodomains to a state that exhibits relaxor behavior.

Ferroelectric versus Relaxor Ferroelectrics

To understand the qualitative features of relaxors, it is useful to compare them with the normal ferroelectrics which are better understood. The distinction between ferroelectrics and relaxor ferroelectrics can be distinguished by three qualitatively different features in the temperature dependence of the dielectric susceptibility. First, in normal ferroelectrics the real part of the dielectric susceptibility shows a Curie-Weiss law behavior at the transition temperature Tc, whereas in the relaxors at the transition temperature the peak has a broad and rounded divergence. This rounded peak position at Tm, marks the dynamic freezing temperature or glass-like transition [13]. Second, there is a strong frequency dependence in the peak position. Last, the polarization in ferroelectrics goes to zero at Tc and in the relaxors the polarization extends well beyond Tm [15]. The fact that there is zero polarization at Tc shows that polar nanodomains vanish, whereas in the relaxor the nanodomains persist well beyond the glassy transition temperature [13].

The normal ferroelectric has a hysteresis loop that at zero field retains large polarization, while in the relaxor this zero field polarization is significantly smaller owing to the fact of the nanodomains being randomly distributed. It is possible to create a large polarization when there is a sufficiently large external electric field. But once the field is removed, polarization returns to being small owing to the randomness of the domains. Finally, it is should be noted that in ferroelectrics, there is a macroscopic structural change at Tc and this occurs in relaxor ferroelectrics [13].

$PbMg_{1/3}Nb_{2/3}O_3$ (PMN) is one of the most extensively studied relaxor ferroelectrics. Different experimental techniques have been used to probe the transition from a glassy state to the long-range ferroelectric ordered state, which included pyroelectric measurements, NMR, dielectric spectroscopy and neutron scattering. For the high temperature and the nature of the phase of this material, the thermal fluctuations at very high temperature were too strong so that there are no well-defined dipoles in the sample. When the temperature is reduced, polar nanodomains begin to form at the Burns temperature (T_d), which is analogous to the spin fluctuations freezing in a magnetic system. The formation of polar nanodomains can be demonstrated by measuring the refractive index n as a function of temperature. Normal perovskite relaxors show deviation from the Curie-Weiss law, which was interpreted that there exist polar nanodomains at up to T_d, upon which they are destroyed and normal Curie-Weiss behavior is obeyed. The measurements of the dielectric susceptibility can provide evidence for the existence of polar nanodomains well above T_m. As temperature decreases, the nanodomains increase in correlation and size, then the departure from Curie-Weiss behavior become more severe [13].

Diffuse Transition

Since polar nanodomains come into existence at T_d, it is important to answer the question: what role do the domains play in describing the relaxor behavior and the apparent glassy nature of the material below T_m? Generally, disorder/randomness plays a crucial role in the transition from the paraelectric state to the glassy transition. There is still much debate about the nature of the diffuse transition of the relaxor ferroelectric in zero electric, which is due to contradictory experimental results. Although there is no general agreement on the nature of state and most of the evidence is in favor of the glassy state behavior, [13] the question to be answered is whether the relaxor behavior is due to a basically glass state with polar nanodomains interacting randomly in the presence of random fields [16], or is the relaxor basically a FE with macroscopic domain broken up into nanodomains subject to quenched random fields, in which the random field interaction causes a smearing out of the FE transitions and lead to relaxor behavior. To answer these remaining questions could promote the further research of relaxors and provide an opportunity for a better understanding of these disorder systems.

2.1.7. Synchrotron Radiation Studies for Relaxor Ferroelectric Crystals - $Pb(Zn_{1/3}Nb_{2/3})O_3$-$PbTiO_3$ (PZNT)

(1) Introduction

Recent developments in regard to relaxor-based single crystal piezoelectrics, such as $Pb(Zn_{1/3}Nb_{2/3})O_3$–$PbTiO_3$ (PZNT), $Pb(Fe_{1/2}Nb_{/1/2})O_3$–$PbTiO_3$ (PFNT) and $Pb(Mg_{1/3}Nb_{2/3})O_3$–$PbTiO_3$ (PMNT) have shown superior performance compared to the conventional Pb(Zr, Ti)O_3 (PZT) ceramics [17, 18]. In particular, their ultrahigh piezoelectric and electromechanical coupling factors in the <001> direction can reach to d_{33}>2000pC/N and k_{33}≈94%, which has attracted tremendous interest and still make these materials very hot.

However, the origin and structural nature of their ultrahigh performance remains an inquisitive but obscure subject of recent scientific interest. To better understand the structural nature of the outstanding properties, it is very important investigating the ferroelectric domain structure in these materials. In ferroelectrics, according to the electrical and mechanical compatibility conditions, domain structures of 180° and non-180° will form with respect to crystal symmetry. There is a close relationship between the domain structure and the crystal symmetry. Through the observation on ferroelectric domain configurations, the crystal structures can be confirmed. Ferroelectric domains are homogenous regions within ferroelectric materials in which polarizations lie along one direction, that influence the piezoelectric and ferroelectric properties of the materials for utilization in memory devices, micro-electromechanical systems, etc. Understanding the role of the domain structure on properties relies on microscopy methods that can inspect the domain configuration and reveal the evolution or the dynamic behavior of domain structure.

It is also well known that the key to solving this issue of exploring the origin of the excellent properties is to reveal the peculiar complex perovskite crystal structures in these materials. Study in structure behavior under high pressure and local structure at atomic level will be helpful for better understanding this problem.

(2) Main Results from the Analysis Methods Based on Synchrotron Radiation

$Pb(Zn_{1/3}Nb_{2/3})O_3$-$PbTiO_3$ (PZNT) and $Pb(Fe_{1/2}Nb_{/1/2})O_3$–$PbTiO_3$ (PFNT) crystal are model ABO_3-type relaxor ferroelectric materials (or ferroelectrics), which demonstrate excellent performance in the fields of dielectrics, piezoelectrics, and electrostriction, etc. To explore the common issues of the structural nature and the relationship between structure and performance of ultrahigh-performance in these materials, in this chapter, the novel X-ray analysis techniques based on synchrotron radiation light, such as synchrotron radiation X-Ray topography, high-pressure in situ synchrotron radiation energy-dispersive diffraction, and XAFS method, are utilized to investigate the domain configuration, structure and their evolution behavior induced by temperature changes and external field.

A. Application of White Beam Synchrotron Radiation X-ray Topography for In-situ Study of Ferroelectric Domain Structures

Ferroelectric domains can be observed by various imaging techniques such as optical microscopy, scanning electron microscopy (SEM), transmission electron microscopy (TEM), X-ray imaging etc. Imaging is normally associated with lenses. Unlike visible light or electrons, however, efficient lenses are not available for hard X-rays, essentially because they interact weakly with matter. Comparatively as an X-ray imaging method, X-ray topography plays a vital role in providing a better understanding of the ferroelectric domain structure. [19] X-rays are more penetrating than photons and electrons, and the advent of synchrotron radiation with good collimation, a continuous spectrum (white beam) and high intensity has given X-ray topography additional powers. The diffraction image contrast in X-ray topographies can be accessed from variations in atomic interplanar spacings or interference effects between X-ray and domain boundaries so that the domain structure can be directly observed (with a micrometer resolution). Especially, via a white beam synchrotron radiation X-ray diffraction topography technique (WBSRT), one can study the dynamic behavior of domain structure and phase evolution in ferroelectric crystals respectively induced by the changes of sample temperature, applied electric field, and other parameter changes.

In this chapter, a brief introduction to principles for studying ferroelectric domain structure by X-ray diffraction imaging techniques is provided. The methods and devices for in-situ studying domain evolution by WBSR are delineated. Several experimental results on the dynamic behavior of domain structure and induced phase transition in ferroelectric crystals accessed at beam line 4W1A of the Beijing Synchrotron Radiation Laboratory (BSRL) are introduced.

Principle of Synchrotron Radiation X-ray Topography

a. X-ray Topography Approach

X-ray diffraction topography is an imaging technique based on Bragg diffraction. In recent decades, X-ray diffraction topography to characterize crystals for the micro electronics industry was developed and completely renewed by the modern synchrotron radiation sources. [19]

Its images (topography) record the intensity profile of a beam of X-rays diffracted by a crystal. A topography thus represents a two-dimensional spatial intensity mapping of reflected X-rays, i.e. the spatial fine structure of a Bragg spot. This intensity mapping reflects the distribution of scattering power inside the crystal; topography therefore reveals the irregularities in a non-ideal crystal lattice. The basic working principle of diffraction topography is as follows: An incident, spatially extended X-ray beam impinges on a sample, as shown in Figure 2.2. The beam may be either monochromatic, or polychromatic (i.e. be composed of a mixture of wavelengths (white beam topography)). Furthermore, the incident beam may be either parallel, consisting only of rays propagating all along nearly the same direction, or divergent/convergent, containing several more strongly different directions of propagation.

A homogeneous sample (with a regular crystal lattice) would yield a homogeneous intensity distribution in the topography (a "flat" image). Intensity modulations (topographic contrast) arise from irregularities in the crystal lattice, originating from various kinds of defects such as cracks, surface scratches, dislocations, grain boundaries, domain walls, etc. Theoretical descriptions of contrast formation in X-ray topography are largely based on the dynamical theory of diffraction. This framework is helpful in the description of many aspects of topographic image formation: entrance of an X-ray wave-field into a crystal, propagation of the wave-field inside the crystal, interaction of wave-field with crystal defects, altering of wave-field propagation by local lattice strains, diffraction, multiple scattering, absorption. Theoretical calculations, and in particular numerical simulations by computer based on this theory, are thus a valuable tool for the interpretation of topographic images. Contrast formation in white beam topography is based on the differences in the X-ray beam intensity diffracted from a distorted region around the defect compared with the intensity diffracted from the perfect crystal region. The image of this distorted region corresponds to an increased intensity (direct image) in the low absorption case.

To conduct a topographic experiment, three groups of instruments are required: an x-ray source, potentially including appropriate x-ray optics; a sample stage with sample manipulator (diffractometer); and a two-dimensionally resolving detector (most often X-ray film or camera). The x-ray beam used for topography is generated by an x-ray source, typically either a laboratory x-ray tube (fixed or rotating) or a synchrotron source. The latter offers advantages due to its higher beam intensity, lower divergence, and its continuous wavelength spectrum. The topography technique combined with a synchrotron source, is well adapted to in-situ experiments, where the material, in an adequate sample environment device, is imaged while an external parameter (temperature, electrical field, and etc.) is changed.

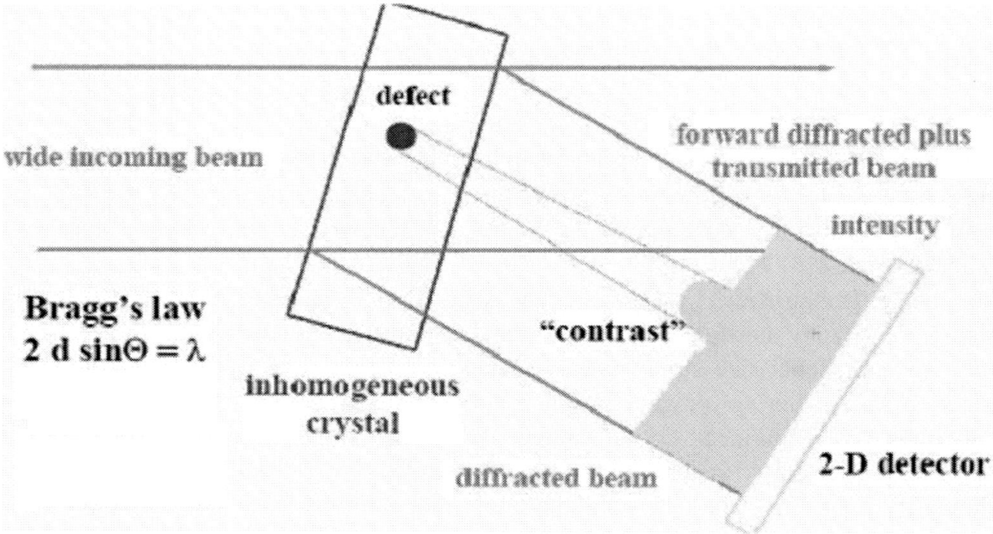

Figure 2.2. The scheme of basic principle of diffraction topography.

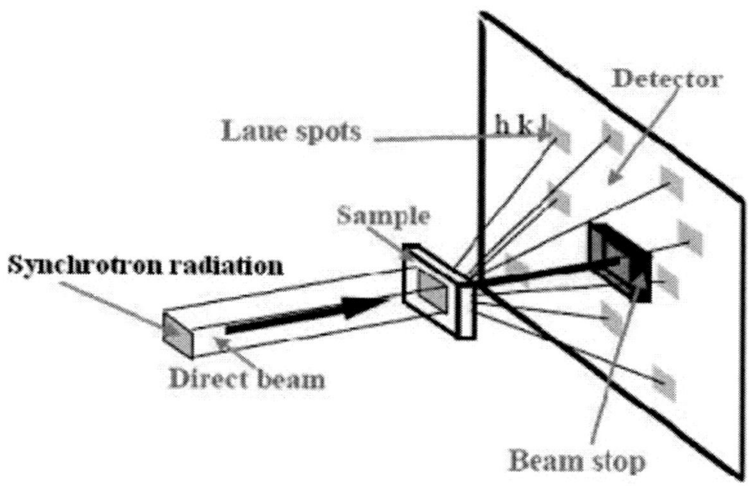

Figure 2.3. Experimental arrangement for synchrotron radiation white beam Laue topography.

b. White Beam X-ray Topography

A simple way to understand the creation of X-ray topographic images is to consider a Laue photograph (Figure 2.3). A polychromatic (white) X-ray beam, containing X-ray energies from about 6 keV to 50 keV (X-ray wavelengths from approximately 2 Å to 0.25 Å), impinges on a crystal. [19] The beam is diffracted in many directions, creating Laue spots. The positions of the diffraction spots appear according to the Bragg equation:

$$E = \frac{hc}{2d\sin\theta_B}, \text{ or } \lambda = 2d\sin\theta_B$$

where E is the incident X-ray energy (and λ is the incident wavelength) selected by crystal planes with spacing d, h is Planck's constant, c is the speed of light, and θ_B is the Bragg angle. Each spot contains uniform intensity if the crystal is perfect.

If, however, the crystal is strained, streaks appear instead of spots due to variations in lattice spacing. In fact, each Laue spot contains a spatial distribution of diffracted intensity attributable to the presence of defects in the crystal. This distributed intensity is difficult to see because Laue spots are typically the same size as the X-ray beam pinhole, and the incident X-ray beam is divergent, but each tiny Laue spot is actually an X-ray topography. At synchrotron radiation facilities, a collimated white X-ray beam can be used to illuminate a sample crystal, and spots with the much larger cross-section of the synchrotron X-ray beam are recorded. The resulting data are an array of Laue spots, as shown in Figure 2.3, each of which is an X-ray topography arising from a different set of atomic planes.

c. Ferroelectric Domain Characterizations

The existence of antiparallel 180° domains is one of the fundamental properties of ferroelectrics. Direct observation of these domains is invaluable to the understanding of the polarization reversal mechanism of ferroelectric structures. Among the techniques of visualizing antiparallel domains, conventional x-ray topography is an important and efficient method although its application is limited by the available characteristic radiations. However, this limitation is easily overcome by white-beam synchrotron radiation topography (WBSRT). A unique aspect of WBSRT is the opportunity that it affords to select out of the continuous spectrum any intended wavelengths to activate strong anomalous scattering. With the WBSRT, it is possible for several diffraction images with anomalous dispersion effect to be activated simultaneously so that the contrast reversal of 180° domains can be directly observed. [20-22] The natural collimation and high intensity of the radiation also make the real-time observation of domain dynamics feasible. As shown in Figure 2.4, when working with a coherent x-ray beam, and if the phases of the structure factors are different, the 180° ferroelectric domains can be revealed.

Considering the mechanical and electrical compatibility conditions, allowed domains in ferroelectrics are the 180o or non-180o ones with the different planes as the domain walls. [21, 22] The extinction condition for a domain wall is:

$$\Delta P \bullet g = 0 \qquad (1)$$

where g is the reciprocal vector of the diffracting plane and $\Delta P = P_1 - P_2$ is the difference of the polarization vectors across the domain wall. The non-180^0 domain structure is illustrated in Figure 2.5.

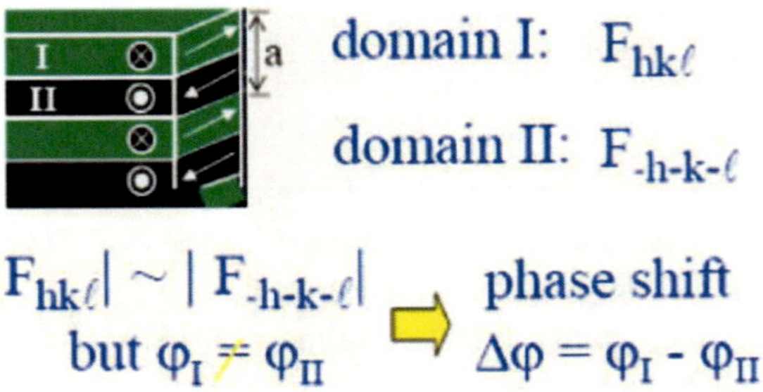

Figure 2.4. (Color on the web only) Scheme of 180° domain.

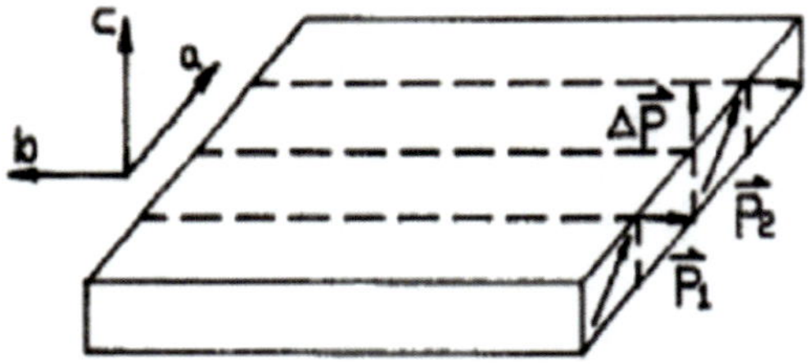

Figure 2.5. Scheme of non-180° domain.

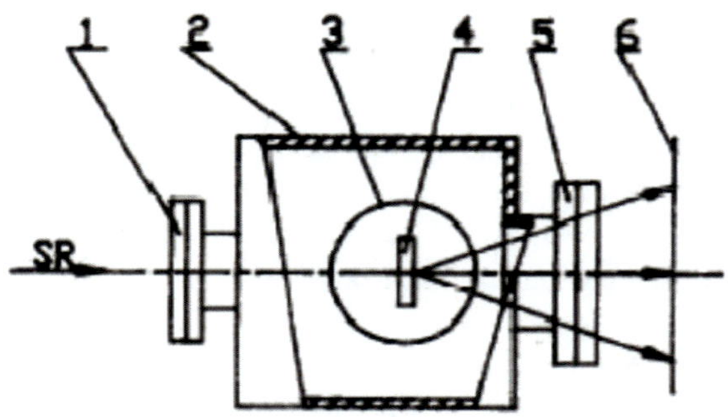

Figure 2.6. Sketch of the high temperature sample chamber: 1-entrance Be window; 2-water cooling environment chamber; 3-furnace; 4-specimen; 5-exit Be window; 6-film.

In-situ Topography Measurements

White beam synchrotron radiation topography not only overcomes the drawback of long exposure time for conventional x-ray topography, but also extends the scope of topography study. The excellent collimation and high intensity of the synchrotron radiation allows the possibility of enlarging the distance between the light source and sample, to improve the image resolution and enlarge the distance between the sample and the detector. This allows the installation of samples inside large environmental chambers with changes of temperature, electric field, or other parameters, to carry out the in-situ topography studies. [23]

a. High Temperature Condition

A high temperature chamber was used for the in-situ topography study. The cylindrical furnace used was made of pure graphite. Two beryllium windows were used for incoming and outgoing x-rays. Two thermocouples attached close to the specimen were used to monitor and control the temperature. A digital control power supply can ramp the electric current smoothly when starting to heat. The temperature control system is based on the Eurotherm controller (model 818) and solid state relay (SSR). By using PID control and time proportion method, the temperature stability is about 0.05□C when holding and 0.1□C when ramping. The on-line PC can set and monitor the temperature via RS-232 interface. A sketch of this chamber is shown in Figure 2.6.

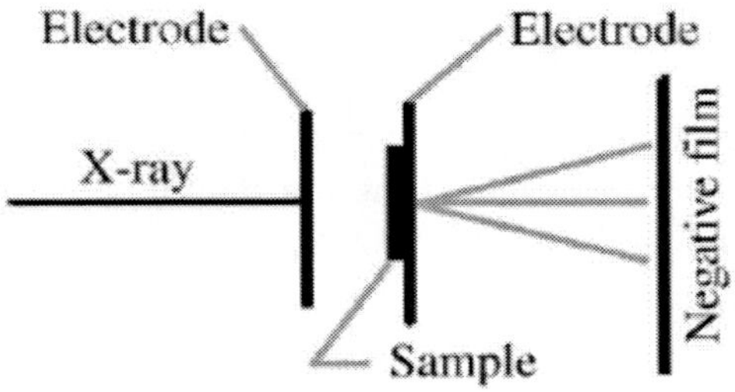

Figure 2.7. The experimental arrangement for applying electric field.

b. DC Electric Field

The change of the ferroelectric domain structure due to the applied DC field can also be observed by white beam synchrotron radiation topography. Figure 2.7 is the experimental arrangement of applying the DC field. The distance between the two electrodes was 4 mm and the applied DC voltage ranged from 0 to 4400 V. The samples can be installed on the surface of plastic plates. The DC electrical field can be applied on two Al electrodes covering on the surface of two X-ray transparent organic materials.

In-situ Topography Study in Ferroelectric Crystals

The in-situ experiments are performed at the Beijing Synchrotron Radiation Facility (BSRF). The x-ray topography station and attached 4W1A beam line are part of the BSRF. The 4W1A is a 45m long white/monochromatic wiggler beam line. The BEPC is operated at the energy of 2.2 GeV and the magnetic field of the wiggler at 1.8T. The topography station situated at the end of the beam line 4W1A is mainly used for the study of the perfection of single crystals, high resolution multi-crystal diffraction and x-ray standing wave research. The main equipment of the station consists of a white radiation topography camera, three environmental sample chambers, an x-ray video imaging system and a four crystal monochromatic camera.

The white radiation topography camera and three environmental sample chambers are used for the dynamic topographic experiments with changes of temperature, stress, electric field or other parameters. The white radiation camera has five axes to rotate the specimen to any orientation with the incident beam and to rotate the detector to collect any diffracted beam.

a. Domain and Temperature-induced Phase Transformation in $0.92Pb(Zn_{1/3}Nb_{2/3})O_3$–$0.08PbTiO_3$ Crystals

The objective of this present work is to investigate temperature-dependence phase evolution in $0.92Pb(Zn_{1/3}Nb_{2/3})O_3$–$0.08PbTiO_3$ (PZN–8% PT) crystals, by employing a real-time white-beam synchrotron X-ray radiation topography method (WBSRT). By combining this technique with other complementary structural experiments, a novel picture of low symmetry phase transformation and phase coexistence is suggested.

PZN–8% PT single crystals used in this experiment were grown by the PbO flux method. A plate perpendicular to [001] axis is cut and well-polished to approximately 200 μm in thickness. Real time observation is performed at the topography station at the 4W1A beam line of the Beijing Synchrotron Radiation Laboratory (BSRL). The storage ring is 2.2 GeV with beam current varied from 50 mA to 90 mA. A cylindrical furnace with coiled heating elements arranged axially around the sample space is used for in-situ topography investigation. After carefully mounting the samples on the hot-stage, we heat them at a slow rate of 0.5 °C/min and observe and record the dynamic topography images by photo films. Through the topography images obtained by this method, we can clearly observe the ferroelectric domain configurations and their evolution as a function of temperature in PZN–8% PT crystals. [24]

Figure 2.8 shows a series of synchrotron radiation topography images with (112) reflection of the (001) crystal plate taken at different temperatures. From Figure 2.8 (a) to (i), we find that the domain structures are very complex. They can be categorized into three kinds of domains and addressed as A, B, and C, as shown in Figure 2.8 (j).

The A domain walls, which are at approximately 45° to the [100] axis, can be obviously observed at room temperature. These domain walls are considered to be the 71° (or 109°) ones in rhombohedral PZN–PT crystals, and can be clearly observed before heating the sample to 132 °C. With increasing temperature from 75 °C to 131 °C, as shown in Figure 2.8 (b)–(c), the B domain laminates become progressively obvious and coexist with the A domains. On the other hand, these domain laminates are along the [010] axis, which can be classified into 90° tetragonal domain walls. At the point of 131 °C, the tetragonal domains become most clear. With heating the sample to above 132 °C, as shown in Figure 2.8 (d), we find that the rhombohedral 71° (or 109°) domain walls (A laminates) become vague, and the image background becomes brighter than before. However, the tetragonal domain walls are still clear. This phenomenon shows that the phase transition from rhombohedral to tetragonal phase (R–T transition) starts at 75 °C, and the tetragonal domains grow gradually.

Most particularly, a set of unique domain walls (C domain walls) appear at this temperature, which is quite different from the A and B domains. This kind of domain wall is shown in Figure 2.8 (k), through the enlarged images taken from 75 °C to 190 °C. As the figures show, the C domain laminates deviate from the (010) direction at 15°–20°. According to the knowledge of domains orientation in crystals with different symmetry and X-ray diffraction extinction relations, as formula (2) shows, these laminates can be considered to be neither rhombohedral nor tetragonal domain structures, but a new monoclinic phase domain structure.

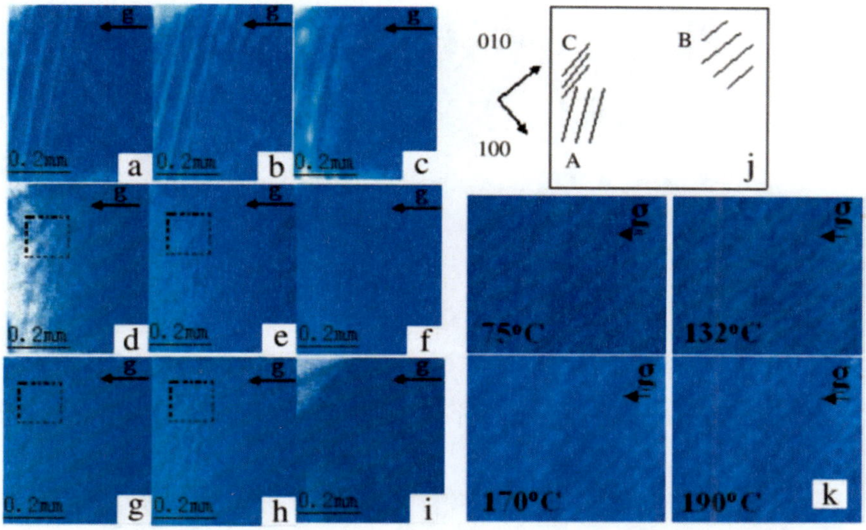

Figure 2.8. (Color on the web only) Images of the in-situ synchrotron radiation topography in PZN–8% PT crystals, the x-rays incident direction to the crystal is [001], the diffraction vector is g = (112): (a) room temperature (20 °C); (b) heating to 75 °C; (c) heating to 131 °C; (d) heating to 132 °C; (e) heating to 190 °C; (f) heating to 262 °C; (g) cooling to 190 °C; (h) cooling to 75 °C; (i) cooling to room temperature; (j) Schematic picture for presenting the ferroelectric domain configurations in the topography images of PZN–8% PT crystals; (k) enlarged images of C domain walls from 75 °C to 190 °C.

With further heating of the system to about 132 °C, we find this domain structure is very stable and coexists with B tetragonal domains. Upon further heating to above Tc (170 °C), the monoclinic C domain structure also remains. This case shows that a monoclinic phase not only appears in the process of ferroelectric–ferroelectric phase transformation, but also coexists with the cubic phase well above TC. With the temperature elevating to about 262 °C, we find nearly all the ferroelectric domains disappear, as shown in Figure 2.8 (f).

When cooling the sample from 262 °C to 75 °C, the monoclinic domains (C laminates), as well as the tetragonal domains (B laminates), are found to reappear, whereas the rhombohedral domains (A laminates) cannot be recovered by cooling to room temperature. During the crystal growth, a rapid cooling process was employed for quick crystallization and to avoid the generation of a pyrochlore phase, which results in a strain field in the crystal. The rhombohedral A domains are expected to be induced by this kind of strain field and preserved at room temperature. Thus, A domains do not reappear after the crystals are re-cooled from 260 °C to room temperature with a slow cooling rate, since this cooling process possibly releases the crystal strain field. However, the monoclinic phase was not generated by the strain field induced during the crystal growth, since the particular C domains as well as B domains can be recovered at 75 °C by slow cooling.

Generally, when the sample temperature reaches the Curie temperature, there occurs a ferroelectric-paraelectric phase transition in conventional ferroelectrics, which results in the disappearance of the domain structure. However in relaxor ferroelectrics, as shown in Figure 2.9 the domain structure can be observed clearly on (111) face of PZN-8%PT crystal when the sample temperature is much higher than the Curie temperature. This phenomenon can be induced by the micro-polarization-region in relaxor crystals.

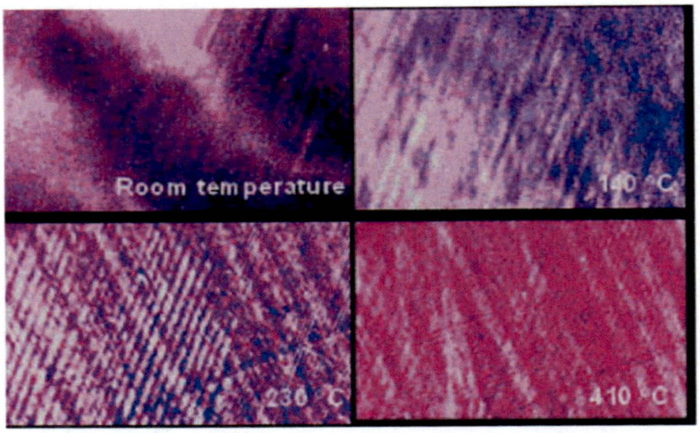

Figure 2.9. (Color on the web only) The temperature induced evolution of the domain structure on (111) face of PZN-8%PT crystal, observed by in-situ synchrotron radiation topography.

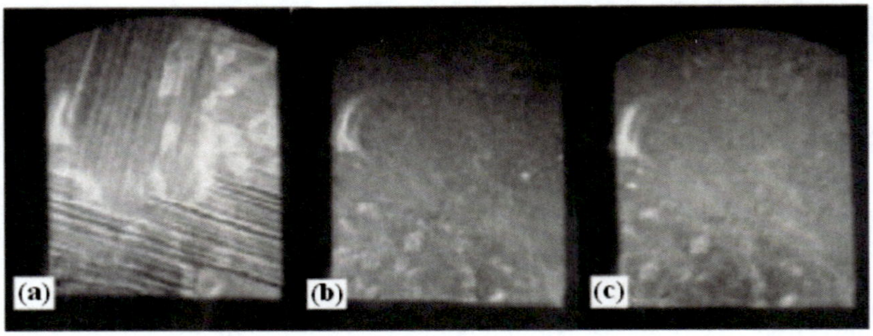

Figure 2.10. The domain structure of GMO varies with the applied DC field.

Through in-situ synchrotron radiation topography under various temperatures, the complex configuration and dynamic evolution of ferroelectric domains in ferroelectric crystals are obtained. It is expected that the present results will encourage more research interest in exploring the origin of the ultra-high piezoelectric and electrostriction properties in ferroelectrics and other advanced materials.

b. Electrical Field Induced Domain Structure

The $Gd_2(MoO_4)_3$ (GMO) crystals were grown by the induction-heated CZ technique. The (0 0 1) crystal pieces of 10–15mm diameter were cut and polished into 2.0 or 0.5mm in thickness. The transparent pieces were using to study the DC electric field induced domain structure by transmission X-ray topography at beam line 4W1A of the Beijing Synchrotron Radiation Laboratory (BSRL) The change of the domain structure due to the applied DC field was also observed. Figure 2.7 is the experimental arrangement of applying the DC field. The distance between the two electrodes was 4mm and the applied DC field ranged from 0 to 4400 V.

Figure 2.10 is a set of series topographs when an applied DC field was added on the GMO crystal piece. Figure 2.10 (a) and (b) are the results obtained when the applied voltage was 600 and 700 V, respectively, and the domain structure did disappear. Figure 10 (c) was the result when the voltage was lowered to zero. From these photographs, we concluded that the multidomain could be transformed to a single domain by applying a DC field and that the single domain could be kept even if the applied DC field was taken away. This experimental result shows us that it is possible to make a periodically poled GMO crystal, despite the presence of ferroelectric and ferroelastic domains in the unpoled GMO crystal. From these results, we concluded that the ferroelectric and ferroelastic multidomains could be transformed to a single domain by the applied DC field, and the single domain could be kept even if the applied DC field was taken away. [25]

B. High-pressure Structural Behavior in PZN-PT Relaxor Ferroelectrics

The ideal structure of ABO_3-type perovskite can be described as a network of corner-linked octahedra, as shown in Figure 2.11. There are B cations at the center of the octahedra and A cations in the space (coordination 12) between the AO_{12} and BO_6 polyhedra. As its B cation is comprised by Zn^{2+}, Nb^{5+}, and Ti^{4+} ions, as well as the A cation comprised by Pb^{2+} ion, PZNT deviates from the ideal perovskite by cation displacements or a rotation (or tilting) of BO_6 octahedra. The possible ''off-centering'' of the B cations makes the crystal structure more complex than that of the ideal ABO_3-type perovskites and reasonably influences the properties. Previously, approaches towards the understanding of the relaxor ferroelectrics were focused on the structural evolution induced by the changes of chemical composition, electrical field or temperature environment. [26-27] These variables induce phase transitions by mainly playing a role of causing displacement of the cation and anion and rotation of BO_6 octahedra in perovskite.

Nevertheless, directly compressing the materials can also induce similar results that will provide a new approach to investigate the complex structure in the ferroelectrics and other functional materials. It is pointed out that the effect of pressure is a ''cleaner'' variable, since it acts only on interatomic interactions. [28] Compared to other parameters, as an extreme variable, high-pressure is of the unique importance in elucidating ferroelectrics, for the unique structural change is susceptible to pressure. Studying the structural changes and compressive behaviors under high-pressure conditions facilitates the understanding of the structural nature of the high-performances in relaxor ferroelectrics or other novel functional materials under normal state.

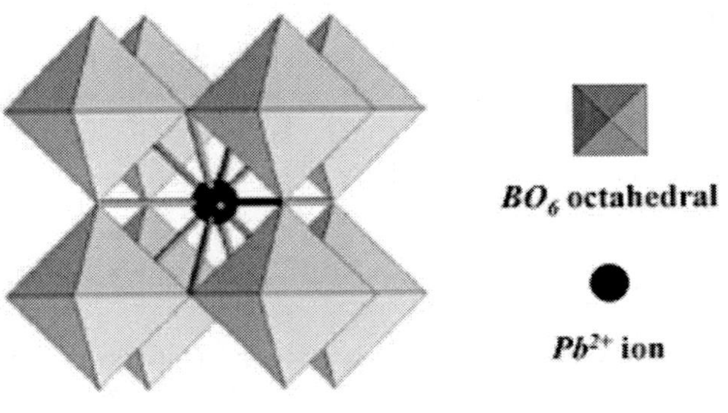

Figure 2.11. Illustration of ideal ABO_3-type perovskite structure, which can be described as a network of corner-linked octahedral.

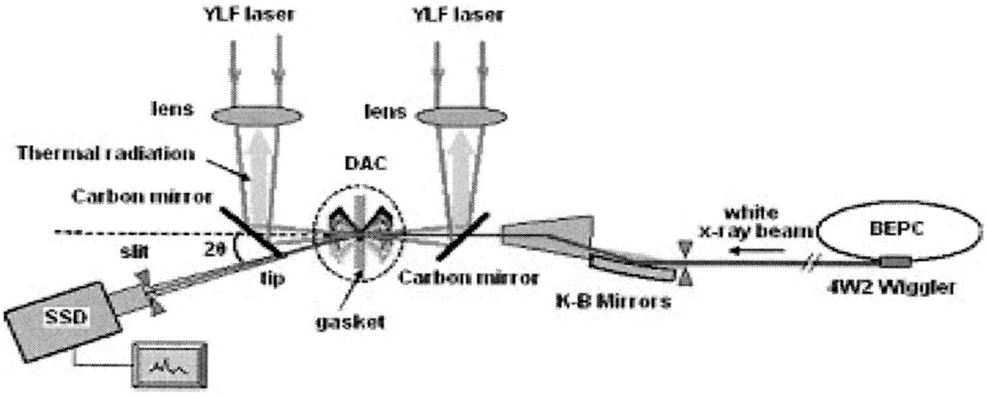

Figure 2.12. The setup of high-pressure X-ray energy-dispersive diffraction experiment.

Thus, recently, the high-pressure structural investigations in relaxor ferroelectrics have become very popular. [29, 30] For instance, Kreisel et al. performed a high-pressure investigation by Ramman spectroscopy of $Pb(Mg_{1/3}Nb_{2/3})O_3$ (PMN), and had obtained the unusual pressure-dependent Ramman relaxor-specific spectra and phase change. By combining the external parameter high pressure with the x-ray diffuse scattering method, a pressure-induced suppression of the diffuse scattering in PMN was indicated. Their observed pressure-induced suppression of diffused scattering above 5GPa is also a general feature in relaxors at high pressure. [31] In this present work, we performed a high-pressure synchrotron radiation energy-dispersive x-ray diffraction (EDXD) investigation on 0.92PZN-0.08PT ferroelectrics, to study the compressive behavior of the materials under high-pressure conditions.

High-pressure Synchrotron Radiation Energy-dispersive X-ray Diffraction (EDXD)

The high-pressure X-ray diffraction patterns employed an energy-dispersive method and were recorded on the wiggler beam line (4W2) of the Beijing Synchrotron Radiation Laboratory (BSRL). [32] A WA66B-type diamond anvil cell (DAC) was driven by an accurately adjustable gear-worm-level system. The 0.92PZN-0.08PT powders and pressure calibrating materials (Platinum powder) were loaded into the sample chamber of a T301 stainless steel gasket. A mixture of methanol/ ethanol 4:1 was used as the pressure medium. Pressure was determined from (111), and (220) peaks of Pt along with its respective equation of state. The *d*-values of the specimens can be obtained according to the energy dispersion equation:

$$E \cdot d = \frac{0.619927}{\sin\theta}(keV \cdot nm) \tag{2}$$

The polychromatic X-ray beam was collimated to a 40×30 μm sized spot with the storage ring operating at 2.2 G eV. The diffracted beam was collected between 5 and 35 k eV and the diffraction 2θ angle between the direct beam and the detector was set at about 15.9552 °. The experiment setup is shown in Figure 2.12. The pressure-induced crystallographic behavior was studied up to 40.73GPa at room temperature in 22 steps during the pressure-increasing process, and the evolution of high-pressure EDXD patterns was obtained.

High-pressure Compressive Behavior in PZNT-PT Crystals

Figure 2.13 shows the selected EDXD patterns of 0.92PZN-0.08PT sample under different pressures, from which the peaks of (110), (111), (200), (210), and (211) indexed in terms of rhombohedral structure can be observed. The strong peaks of (111) and (200) of Pt, whose photonic energy is 19.72keV and 22.77keV respectively, will also be clearly observed. Apart from these diffraction lines, several fluorescence peaks of Pb and Nb emerged in the low-energy side of the curve.

With increasing pressures from the ambient pressure (0 GPa), the (110) and (200) peak become broader and the intensity of (210) and (211) peaks start to decrease from 5.17GPa. On further increasing pressure, the intensity of (210) and (211) peaks become weaker from 21.34GPa, and the (210) peak vanishes at the pressure of 28.38 G Pa. at about 40.73GPa, and these two peaks seem to vanish. The abrupt changes of the EDXD patterns indicates that the phase transitions can possibly be induced by applying pressures, and the estimated transition pressure point is at about 5.17GPa and 28.38GPa, respectively.

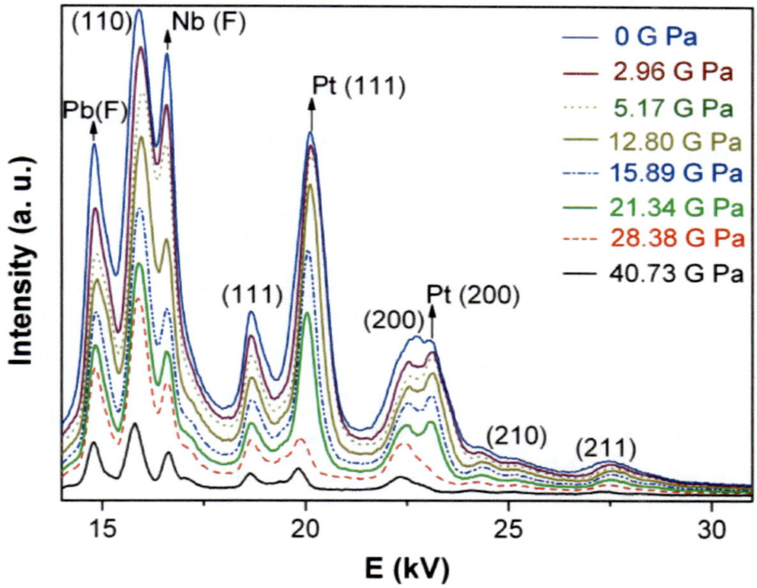

Figure 2.13. (Color on the web only)The EDXD curves of PZNT sample under different pressures, the maximum pressure applied here is 40.73 G Pa. Pb (F) and Nb (F) indicate the fluorescence peaks of Pb and Nb emerged in the low-energy side.

In order to better understand the pressure-induced phase transition and compressive behavior of 0.92PZN-0.08PT, we calculated the structure parameters such as interplanar spacing value (d-value) evolving under various pressures according to experiment results and equation (1), as shown in Figure 2.14. It is noticed that, as a function of pressure, the d-value decreasing issues of different peaks of (110), (200), and (211) are greatly different, which reveals that the compressibility of structural parameters is anisotropic. To investigate the compression behavior of the sample in different pressures, we divided the $d_{(110)}$ vs. pressure curve into several zones. It shows the decreasing rate of $d_{(110)}$ spacing in different pressure zones is different, i.e., the decreasing rate of the zones of *A*, *C*, *E*, and *G* are not identical, which should demonstrate that the change of compressibility is also nonlinear.

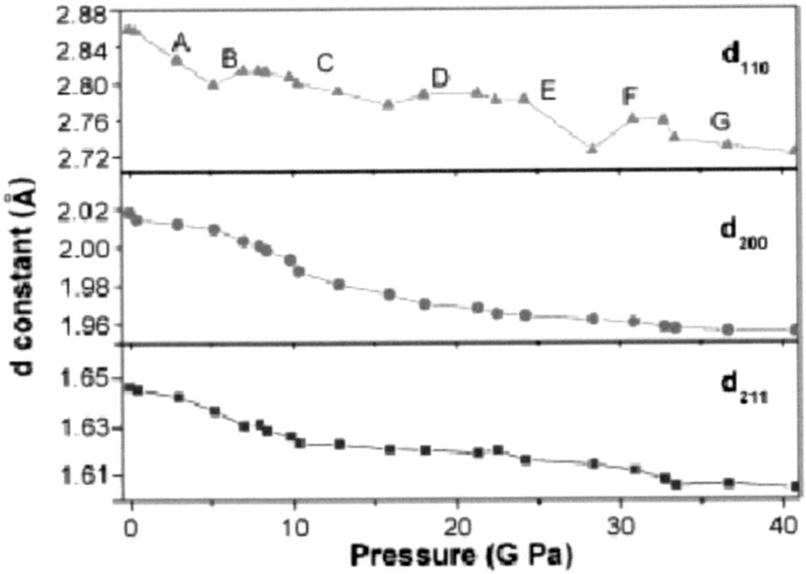

Figure 2.14. The curves of d-spacing parameters of different diffraction peaks under various pressures. Top part: $d_{(110)}$ vs. pressure; Middle part: $d_{(200)}$ vs. pressure; and Bottom part: $d_{(211)}$ vs. pressure.

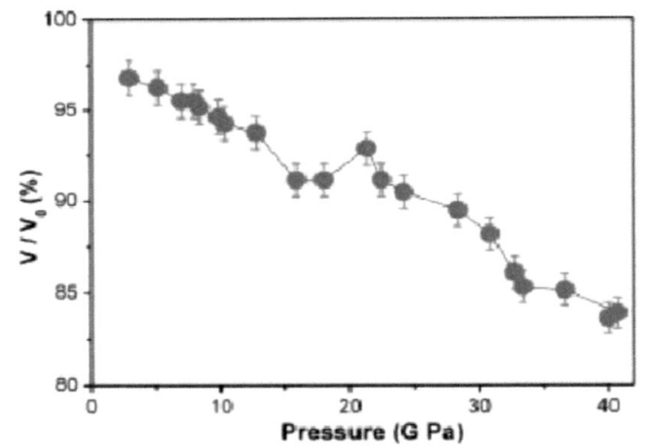

Figure 2.15. The curves of the relative volume V/V_0 vs. pressure curve of 0.92PZN-0.08PT sample.

The abrupt changes appeared in the *B*, *D*, and *F* zones of the curve, which are correspond to the three pressure ranges of about 5.17-7.5GPa, 15.2GPa-21.4GPa, and 30.3-34.5GPa, respectively, showing several kinky ranges. [33] We name this phenomenon "multi-kinky" pressure behavior. This behavior is considered not to be induced by the measurement errors according to the error analysis. Generally, the energy resolution of the detector in EDXD experiment is a key parameter to determine the diffraction peak position. In the present experiment, the Si (Li) detector is adopted and its energy resolution is estimated to be about 170 e V, from which the measuring error of diffraction peak position may be calculated to be less than 3 eV, corresponding to a relative measuring error of structural parameter of $(\Delta d/d) \approx$

10^{-4}. For instance, in Figure 13a, the calculated ($\Delta d/d$) of the (110) face in kinky ranges is estimated to be at the order of 10^{-2}. Therefore, the abnormal change of *d*-spacing under high-pressure should be intimately related to the structural characteristics of ABO_3 compounds. Particularly, in the *B* and *D* zones, the $d_{(110)}$ value first increases and then decreases with the applied pressures, which seems to show an abnormal compressive behavior. While in *F* zone, the $d_{(110)}$ value decreases with the increased pressure. This behavior can be attributed to the abruptly changed strain induced by structural transformation from phase *E* to the phase *G*. The *B*, *D*, and *F* zones can also be considered as the areas for phase-coexistence, or transformation regions. [34] The d-spacing vs. pressure curves of (200) and (211) peaks also exhibit multi-kinky changes and nonlinear compressive behavior. Therefore, it is estimated that this multi-kinky behavior is indicative of pressure-induced multi-phase transition and phase coexistence occurs in 0.92PZN-0.08PT crystals. The accurate pressure-induced lattice changes extracted from the high-pressure EDXD experiments will be published elsewhere.

Figure 2.15 shows the relative volume V/V_0 vs. pressure curve of 0.92PZN-0.08PT, from which can be seen that the volume of the unit cell changes with the increasing pressure. Similar to the curve of $d_{(110)}$ vs. pressure, this curve can also demonstrate interesting multi-kinky compressive behavior. We roughly fit the experimental data to Birch-Murnaghan (BM) equation and get the equation of state (EOS) of 0.92PZN-0.08PT as following:

$$P = \frac{3}{2}B_0\left[\left(\frac{V}{V_0}\right)^{-\frac{7}{3}} - \left(\frac{V}{V_0}\right)^{-\frac{5}{3}}\right] * \left\{1 - \frac{3}{4}\left(4 - B_0^1\right) \times \left[\left(\frac{V}{V_0}\right)^{-\frac{2}{3}} - 1\right]\right\} \quad (4)$$

the bulk modulus B_0 and its first-order derivative B_0^1 for 0.92PZN-0.08PT system are obtained as $B_0 = 267 \pm 15$GPa, $B_0^1 = 4$.

As shown in Figure 2.14 and Figure 2.15, the structure parameters (d and V) display several abrupt changes around several crossover pressures (or during several pressure ranges), which are suggestive of the anomalous phase transitions and multi-kinky compressive behavior of 0.92PZN-0.08PT crystals under high-pressure conditions. It is speculated that this kinky behavior of the structural parameters of 0.92PZN-0.08PT could also be a more general feature in ABO_3-type perovskite oxides. For instance, some similar anomalous behaviors had also been presented in $BaTiO_3$ and $PbTiO_3$. [20] Through the calculations using a first-principles approach based on density-functional theory, an enormous tetragonal strain can be induced in these two simple perovskites by the application of negative hydrostatic pressure. Their structural parameters such as cell volume and atomic displacements were found to display a kinky behavior suggestive of proximity to a phase transition. Comparatively, in the present work, a multi-kinky compressive behavior in 0.92PZN-0.08PT which is more complex than single kinky behavior appeared in $BaTiO_3$ and $PbTiO_3$ was obtained. In a way, the different compressive behavior between 0.92PZN-0.08PT and the simple peroveskites may be ascribed to the chemical composition variation. The *B* cation of $BaTiO_3$ and $PbTiO_3$ is only comprised by the Ti^{4+} ion, while that of 0.92PZN-0.08PT is comprised by Zn^{2+}, Nb^{5+}, and Ti^{4+} ions, which makes the structure of 0.92PZN-0.08PT more complex than that of $BaTiO_3$ and $PbTiO_3$.

The unique nonlinear behavior in 0.92PZN-0.08PT can also be explained in terms of the relative compressibilities of the octahedral (BO_6 polyhedral) and dodecahedral (AO_{12}

polyhedral) cation sites in the perovskites structure, or of the compressibilities of the A-O and B-O bonds, as shown in Figure 2.11. It is clear that the relative compressibilities of the AO_{12} and BO_6 sites must play an important role in determining whether the perovskite structure becomes more or less distorted with increasing pressure. [22] From the curves of cell volume or d-spacing v.s pressure, we found that the bulk compressibilities in 0.92PZN-0.08PT varying with applying pressures. Since 0.92PZN-0.08PT is of perovskite structure comprised by AO_{12} and BO_6 polyhedra, we think the evolution of the bulk compressibilities in it can also be ascribed to the relative compressibilities of the AO_{12} and BO_6.

The so-called polyhedral approach is based on the observation that the identification of cation–oxygen polyhedra, M_xO_y, not only simplifies the description of complex crystal structures, but also allows a better understanding of physical properties for materials containing similar polyhedra. [23, 24] Considering the influence of the bond length and the anion/cation charge of a given polyhedron on the compressibility, the empirical expression for the compressibility of a given polyhedron were derived:

$$K_{poly} = \frac{4}{3}\frac{d^3}{S^2 Z_0 Z_a} GPa^{-1} \tag{5}$$

Z_a is the anion charge, Z_0 the cation charge, d the bond length in Å and S^2 an ionicity scaling factor equal to 1/2 for oxygen-based polyhedra. The extension of this approach is to assume that the compressibility of complex crystals with linked polyhedra can be derived when κ_{poly} for each polyhedron is known. Based on the consideration of the influence of cation–anion distances and angles between polyhedra (polyhedra tilting), J. Kresiel et al proposed a simple model to deduce the bulk compressibility, κ_{bulk}, from the compressibility of each constituent polyhedron:

$$K_{bulk} = \sum_i x_i \frac{V_i}{V_0} k_i^{poly} \tag{6}$$

where the index i refers to a given type of cation–anion polyhedron and x_i is the concentration on a crystallographic site. V_i and V_0 are the volumes of a given polyhedron and the unit cell, as described by

$$\sum_i \left(x_i \frac{V_i}{V_0} \right) = 1 \tag{7}$$

Since in the PZN-PT crystal, the AO_{12} polyhedron is PbO_{12} dodecahedra and the BO_6 polyhedra is comprised by Zn-O group, Nb-O group, and Ti- O group. The equation (6) was modified to estimate the compressive behavior of 0.92PZN-0.08PT as following:

$$K_{PZNT} = \frac{V_{PbO12}}{V_{PZNT}} + x\frac{V_{ZnO6}}{V_{PZNT}} + y\frac{V_{NbO6}}{V_{PZNT}} + z\frac{V_{TiO6}}{V_{PZNT}} \tag{8}$$

where x, y, z implies the concentration of Pb^{2+}, Zn^{2+}, Nb^{5+}, and Ti^{4+} ions, respectively. Due to these cations being chemically very different, it is believed that the compessibilities of these MO_6 polyhedra will be different on applying various pressures, which influence the total compressibilities in 0.92PZN-0.08PT crystals. Therefore, although this high-pressure diffraction technique only allows the determination of the mean bond distance, *d*, averaged over the entire crystal for each distinct polyhedron, the modified model can be considered as an approach to explain the compressive behavior and the change of compressibilities of 0.92PZN-0.08PT crystal under high-pressure conditions.

Therefore, through the method of high-pressure synchrotron radiation energy-dispersive x-ray diffraction, the abnormal compressive behavior in 0.92PZN-0.08PT under high-pressure has been observed, which is considered to be intimately related to the structural characteristics of 0.92PZN-0.08PT crystals. The high-pressure compressive behavior reflects and uncovers abundant structural information under extreme conditions, which is helpful to intrigue significant interests in exploring the structural nature and chemical (or physical) origin of ultrahigh performance in relaxor ferroelectric materials and other functional materials.

C. XAFS Study in Relaxor Ferroelectrics

Relaxor ferroelectrics are a class of ferroelectrics that have a diffuse, frequency-dependent permittivity maximum, with a relaxation spectrum much broader than the Debye-type. But these features do not provide a direct link to the key structural element that is essential for relaxor ferroelectrics. A common, crucial element in all Pb containing relaxors, Pb(B', B'')O3, is a size mismatch between the ferroelectrically active B'' cations and the ferroelectrically inactive B' cations. The data of extended X-ray absorption fine structure (EXAFS) based on synchrotron radiation can provide local structural evidence in indicating the structural origin of the anomalous properties in relaxors.

Synchrotron Radiation XAFS Approach

X-ray absorption fine structure (XAFS) refers to the details of how x-rays are absorbed by an atom at energies near and above the core-level binding energies of that atom, as shown in Figure 2.16. Specifically, XAFS is the modulation of an atom's x-ray absorption probability due to the chemical and physical state of the atom. XAFS spectra are especially sensitive to the formal oxidation state, coordination chemistry, and the distances, coordination number and species of the atoms immediately surrounding the selected element. Because of this dependence, XAFS provides a practical and simple way to determine the chemical state and local atomic structure for a selected atomic species. XAFS can be used in a variety of systems and bulk physical environment.

X-ray absorption measurements are relatively straightforward, provided one has an intense and energy-tunable source of x-rays, such as a synchrotron radiation source. Many experimental techniques and sample conditions are available for XAFS, including such possibilities as very fast measurements of high spatial resolution. Since the characteristics of synchrotron sources and experimental station dictate what energy ranges, beam sizes, and intensities are available, this often puts practical experimental limits on the XAFS measurements that can be done even if there are few inherent limits on XAFS.

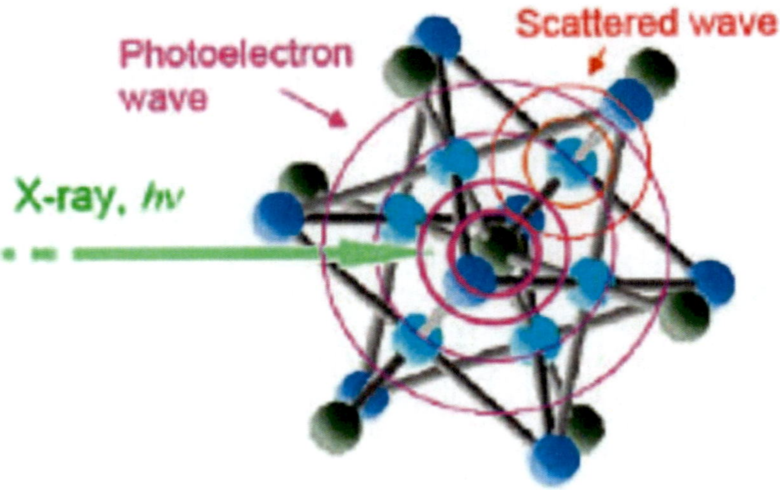

Figure 2.16. (Color on the web only) Illumination of XAFS measurements.

The x-ray absorption spectrum is typically divided into two regimes: x-ray absorption near-edge spectroscopy (XANES) and extended x-ray absorption fine-structure spectroscopy (EXAFS). Though the two have the same physical origin, this distinction is convenient for the interpretation. XANES is strongly sensitive to the formal oxidation state and coordination chemistry (e.g., octahedral, tetrahedral coordination) of the absorbing atom, while the EXAFS is used to determine the distances, coordination number, and species of the neighbors of the absorbing atom.

XAFS Study in $Pb(Fe_{1/2}Nb_{1/2})_{1-x}Ti_xO_3$ (PFNT) Relaxor Crystals

Lead iron niobate $Pb(Fe_{1/2}Nb_{1/2})O_3$ (PFN) has been studied with great interest recently because of its multiferroic and relaxor ferroelectric properties. $Pb(Fe_{1/2}Nb_{1/2})_{1-x}Ti_xO_3$ (PFNT) is a modified material which has also been studied extensively in recent years. PFNT undergoes a rhombohedral-tetragonal phase transition with the increase of the Ti concentration. [38] Recent study reveals that the morphotropic phase boundary (MPB) is in the range $0.06 < x < 0.08$. [39] The PFNT ceramics (x=0.13) shows a diffuse phase transition at 398 K. [40] The (001)-cut PFNT (x=0.48) single crystal undergoes a tetragonal-cubic phase transition at 518 K, but no Curie temperature was detected below 570 K in the (001)-cut PFNT (x=0.06) crystal. [41, 42]

We investigated the dielectric properties of PFNT single crystals below room temperature. [43] The EXAFS results provided a clear picture of the local structure of iron ions in PFNT, and the pre-edge XAFS suggests that the observed dielectric anomaly is induced by the hopping conductivity between Fe^{2+} and Fe^{3+}. We report the low temperature (90~300 K) dielectric properties of the (100)-cut PFNT single crystals. Two dielectric anomalies induced by different mechanisms were observed in the low Ti-doped sample, while only one was detected in the high Ti-doped sample. The mechanism of the dielectric anomalies were discussed based on the microstructures investigated by X ray absorption fine structure (XAFS) results.

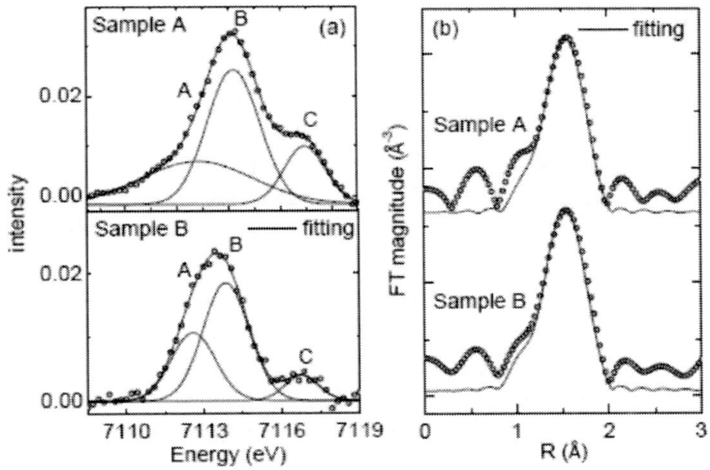

Figure 2.17. (a) Pre-edge absorption spectra of Fe K edge, symbols are experimental data, and lines are fitting results by Gaussian function; (b) Fourier transform of Fe K edge (first coordinate shell), symbols are experimental data, and lines are fitting results.

High quality PFNT single crystals (x=0.07, named sample A; x=0.48, named sample B) were fabricated without any impure phase. Both samples were cut with the surface parallel to (100) plane. Pre-edge and extended x ray absorption fine structure (EXAFS) data were collected at the beam line 13B of Photo Factory, National High Energy Institute of Japan at room temperature. To measure the dielectric properties, aluminum electrodes were evaporated on both sides of the samples. [43]

When considering activation energy, it seems that the A-site doping can change the activation energy values but the B-site doping can not. Considering the different B-site doping of our two samples, though the crystal structures are very different, they have similar activation energy values. [43] Figure 2.17 (a) shows the background-subtracted pre-edge XAFS spectra at Fe K-edge for both samples, with tree peaks (named A, B and C). The experiment data is fit perfectly using three Gaussian functions. The positions of the three peaks are determined to be 7112.8, 7114.2 and 7116.9 eV for sample A, and 7112.6, 7113.9 and 7116.8 eV for sample B, respectively. All the three peaks of sample B are slightly lower energy shifted than sample A, approximately 0.1~0.3 eV. This might be due to their different crystal structures. Peak B is located around 7114 eV, which is close to the value reported for ferric compounds. The separation between peak A and peak B corresponds to the typical separation of the order of 1.4 eV that has been reported for Fe2+ and Fe3+, therefore, Fe2+ ions exist in our crystals. [44] The pre-edge XAFS suggests that the first anomaly is induced by the hopping conductivity between Fe2+ and Fe3+. The local structure of iron ions is studied by EXAFS at Fe K-edge. Figure 2.17(b) shows the Fourier transform of the Fe K-edge spectra for both samples. The first coordinate shell (Fe-O) is well separated and analyzed, and the structure parameters are listed in Table I.

The coordinate number of both samples is smaller than 6, implying that there are oxygen vacancies in our crystals. Moreover, the sample A has a smaller coordinate number than the sample B, indicating a higher oxygen vacancy concentration of sample A than sample B, which can be also confirmed by the impedance spectra (not shown here) that sample A has much smaller resistance than sample B. The distance of Fe-O for both samples is very close.

It is worth noticing that the two samples have a close Debye-Waller (DW) factor value, implying that the local environment of iron ions in sample A is similar to that in sample B. Based on the EXAFS results, we conclude that the B-site doping hardly changes the local structure of iron ions. Thus, the electrons might need the similar energy to overcome the potential barrier to hop between Fe2+ and Fe3+, leading to the close activation energy value. On the other hand, we assume that for the A-site doping, the local structure of iron ions might be distorted significantly, resulting in the different activation energy values.

Table 1. (Color on the web only) Parameters of the first coordinate shell of Fe K edge: N is coordinate number, R is the distance of Fe-O, σ2 is Debye-Waller factor, E0 is energy shift, and R is relative error

x	N	R	σ^2 (10^{-3})	E_0	R factor (%)
0.07	5.4±0.2	2.011±0.003	6	-5.5±0.5	1.1
0.48	5.7±0.1	2.007±0.005	6	-8.2±0.8	2.3

CONCLUSION

The principles for studying ferroelectric domain structure by white beam synchrotron radiation X-ray diffraction topography (WBSRT)has been introduced, and several results on the dynamic behavior of domain structure and phase transition in several ferroelectric crystals induced by temperature and DC electric field had been introduced. Due to its simple principle and accessible setup, WBSRT has become a unique method to characterize ferroelectric materials. WBSRT offers an elegant solution to observe ferroelectric domain structures, to study the dynamic domain behavior and in-situ phase evolution induced by the changes of temperature, electric field, or other parameters, along with other structural study techniques it is possible to give insight into various phenomena occurring in ferroic materials. Through the method of high-pressure synchrotron radiation energy-dispersive x-ray diffraction, the abnormal compressive behavior in 0.92PZN-0.08PT under high-pressure have been observed, which is considered to be closely related to the structural characteristics of 0.92PZN-0.08PT crystals. The high-pressure compressive behavior reflects and uncovers abundant structural information under extreme conditions. The XAFS did a good job of providing a clear picture of the localized state hopping conductivity, and the similar local environment of iron ions is the origin of the close activation energy of the samples. All of these findings will be helpful to stimulate interest in exploring the structural nature and chemical (or physical) origin of ultrahigh performance in relaxor ferroelectric materials and other functional materials.

2.2. High Curie Temperature Ferroelectrics: Simple Overview

Many oxide materials with the perovskite structure, such as Pb(Zr1-xTix)O3 (x~0.48) (PZT) system, having the general formula ABO_3, form the backbone of the ferroelectrics industry. These materials have come into widespread use in applications that range in sophistication from medical ultrasound and underwater sonar systems to relatively mundane devices to novel applications in active and passive damping systems for sporting goods and automobiles [45-49].

Recent developments in regard to relaxor-based single crystal piezoelectrics, such as $Pb(Zn_{1/3}Nb_{2/3})O_3$– $PbTiO_3$ (PZNT) and $Pb(Mg_{1/3}Nb_{2/3})O_3$– $PbTiO_3$ (PMNT) have shown superior performance characteristics compared to Pb(Zr, Ti)O_3 (PZT) polycrystalline ceramics [50-53]. In particular, their ultrahigh piezoelectric and electromechanical coupling factors in the <001>direction can reach to d_{33}>2000pC/N and k_{33}≈94%, which have attracted tremendous interest and stimulated extensive research. The enhanced piezoelectric activity of PZNT and PMNT, however, comes with the sacrifice of the temperature stability of the properties, being limited to their relatively low Curie temperatures (T_C≈178 °C), and further restricted by their lower ferroelectric phase transition temperatures (*Tr-t*≈ 120 °C). In some important applications, especially in the automotive and aerospace industries, the need for actuation and sensing over a broad temperature range is essential. Therefore, the low Curie temperature and relatively low ferroelectric phase transition temperatures (rhombohedral to tetragonal phase) of PZNT and PMNT systems limit their implementation in many fields. On the other hand, the temperature stability of the electrical properties, as well as good ferroelectric or piezoelectric activity, is important for device applications. So, ferroelectrics with high transition temperatures (*Tc)* are attractive [54-57].

Many relaxor-PbTiO3 ferroelectrics, such as $Pb(In_{1/2}Nb_{1/2})O_3$-$PbTiO_3$(PINT), $Pb(Yb_{1/2}Nb_{1/2})O_3$-$PbTiO_3$ (PYNT) and $BiScO_3$-$PbTiO_3$ (BSPT) systems, have Curie temperatures as high as that of PZTs's or about 100°C higher than that of PZTs, and hold excellent properties, which can overcome the shortage of PZNT and PMNT crystals, and meet the requirements in applications as next generation materials for actuators in automotive and aerospace sensors.

The BS-PT system is based on the general solid solution formula (1-x)$BiMeO_3$-x$PbTiO_3$, where Me is a suitably sized 3^+cation (Sc, In, Yb, etc.). Based on the perovskite formula ABO_3 and the geometrical packing of charge spheres, a concept of tolerance factor (t) was proposed as

$$t = \frac{r_A + r_o}{\sqrt{2}(r_B + r_o)} \quad (9)$$

where r_A, r_B, and r_O are the respective ionic radii [58-59]. The ideal tolerance factor for a perovskite material is equal to 1. Based on the tolerance trends in perovskite ferroelectrics, a relationship between tolerance factor (t) and Curie temperature(Tc) was proposed, namely, the tolerance factor can be used to estimate the changes of Curie temperature with different perovskite systems [44, 60, 61].

From the empirical analysis, $BiMeO_3$-based perovskites with decreasing values (t) were suggested to confirm the premise that Tc increases with decreasing tolerance factor. It is evident that a solid solution of $BiScO_3$(t=0.907), $BiInO_3$(t=0.884), or $BiYbO_3$(t=0.857) with $PbTiO_3$ would exhibit MPBs with Tc that were significantly greater than PZT(t=0.96, Tc=386°C). Solid solutions with $BiMe^{+3}O_3$-$PbTiO_3$ have small tolerance factors and are proving to give new high temperature ferroelectric solid solutions. Although, the Tc of $BiInO_3$-$PbTiO_3$ and $BiYbO_3$-$PbTiO_3$ systems may reach to 560°C and 700°C, respectively, the instability of the peroveskite phase in these two materials prevents them from being easily synthesized. For (1-x)$BiInO_3$-x$PbTiO_3$, pure perovskite phases could not be achieved even for x≧0.9. To enhance the perovskite stability, the addition of 1-3 mole% $BaTiO_3$ was added into the system for the compositions with x =0.7 [44]. Although the existence of a MPB was not confirmed, the dielectric temperature behavior for a tetragonal composition (x=0.7) revealed a Tc = 570°C, being significantly higher than that of PZT and $PbTiO_3$ themselves. The transition temperature-tolerance factor relationship was further substantiated in the (1-x)$BiYbO_3$-x$PbTiO_3$ system, although the sample was a mixed phase. For the high Tc ferroelectric systems, the (1-x)BiScO3-x$PbTiO_3$ system exhibits a MPBs and having a Tc greater than 450°C.

Recently, the (1-x)$BiScO_3$-x$PbTiO_3$ solid solution system has demonstrated the potential for producing piezoelectrics with higher piezoelectric activity, higher in-use temperatures, and lower temperature-property dependence than any present-day ferroelectric material, with the perovskite or any other crystal structure [62-65].

$BiScO_3$ (BS) ceramics were first synthesized in 1969 [66], under a high pressure by Tomashpol'skii et al. Under ordinary pressure, pure $BiScO_3$ is unstable. But (1-*x*)$BiScO_3$-*x*$PbTiO_3$ (BS-PT) exhibits a temperature stability of the electrical properties until 350 °C and has a T_C of about 450 °C. The phase diagram of BS-PT system indicates that, at room temperature, the rhombohedral symmetry lies in the region of 50<x<62mol% $PbTiO_3$ (PT), and the tetragonal phase lies in the range x>65mol% PT. In the solid-solution range 62<x<65mol% PT, there exists a MPB phase [45, 67].

BS-PT ceramics exhibit good piezoelectric properties and high ferroelectric transition temperature comparable to PZT ceramics near the MPB. Using conventional solid state processing techniques, rhombohedral, MPB, and tetragonal phase BSPT ceramics (at approximately 62-65%$PbTiO_3$) had been prepared respectively [45, 65, 68, 69]. In the composition of x=0.64, especially, high piezoelectric properties, d_{33}=460pC/N and high curie temperature, Tc=450°C is obtained. The results of high temperature piezoelectric measurements show high piezoelectric coefficients persist to temperatures approaching 400°C.

In order to improve the piezoelectric properties of the BS-PT system, Zhang et al [54 70-73] performed the investigations on growth and characterization of BS-PT single crystals. They focused their attention on the composition of 0.43$BiScO_3$– 0.57$PbTiO_3$. BiScO3-PbTiO3 (BSPT) single crystals have been grown by the flux method. The piezoelectric properties of (001) 0.43BiScO3– 0.57PbTiO3 single crystals can be found within the papers of Zhang SJ:

(1) electromechanical coupling factor (*k*33) about 90% at room temperature;
(2) piezoelectric coefficient over 400 pC/N at room temperature;

(3) Curie temperature is 402-436 °C;
(4) high coercive field is 13.7-28 kV/cm, or kV/cm;
(5) the crystal exhibited a very good temperature stability of the properties till 380 °C;
(6) a rhombohedral–tetragonal phase transition temperature (*Tr-t*) around 350 °C;

The piezoelectric properties of BS-PT crystals can reach the order of those of relaxor-$PbTiO_3$ single crystals, but their usage temperature range for their application is broader. The enhanced properties in the BS-PT crystal make this material a promising candidate for high temperature and temperature stable actuators and transducers. However, it is difficult to grow large size and good quality single crystals, and the cost of crystal growth is considerably higher than that of growing ceramics and thin films.

It is important to notice that the same properties remain preserved also in BS-PT thin films [68]. Since temperature stability of the electric properties, as well as good ferroelectric or piezoelectric activity, is important for MEMS device application, ferroelectric thin film with high transition temperatures are attractive. Micro-Electro-Mechanical Systems (MEMS) are the integration of mechanical elements, sensors, actuators, and electronics on a common silicon substrate through microfabrication technology. Electromechanical actuators directly transform input electrical energy into mechanical energy. T. Yoshimura et al [74] reported that 1-μm thick epitaxial films of (001) $BiScO_3$-$PbTiO_3$ near MPB had been grown on (100) $SrRuO_3$/(100) $LaAlO_3$ substrates by pulsed laser deposition from a $BiScO_3$-$PbTiO_3$ (40/60 or 50/50) ceramic target. The films had room temperature dielectric constant of 850-800, $\tan\delta$ = 0.08-0.09, and maximum dielectric constant of 5530-3600 at 455-460 °C. The remanent polarization is about 42-36 μC/cm^2 and the coercive field was 75 kV/cm. The effective transverse piezoelectric coefficient e_{31f} was –9 to -12 C/m^2. This result is very encouraging for sensor and actuator device development because the observed piezoelectric properties are as good as (001) oriented Pb(Zr, T)O_3 films ($e_{31,\ f}$ ~ -12 C/m^2), while the transition temperature is 100 °C higher than that of PZTs.

Performance of High Temperature Piezoelectrics

For instance, in the application of actuators, an applied voltage has to be maintained across the material to control the magnitude of the actuation. For a given applied field, the excellent performing actuator materials produce a large strain, implying that they have a high piezoelectric 'd' coefficient. As temperature increases and approaches T_C, the piezoelectric charge coefficient also increases. This allows dipoles to move more easily, but also leads to rapid thermal depoling and a dramatic reduction in the effective d coefficient, with resulting increased power consumption. Generally, this thermal depoling and power consumption limit the temperature usage to less than half of T_C. In practice, the magnitude of piezoelectric coefficients should balance with an appropriate temperature dependence to maximize device performance.

2.3. Current State-of-the-Art

Piezoelectric materials lead to a wide variety of commercially interesting applications. There are two general classes of these materials: non-MPB and MPB based materials.

Non-MPB piezoelectric materials generally exhibit low values of relative permittivity and small piezoelectric coefficients. Many these materials are based on the BO_6 octahedra, which are generally the building blocks of the perovskite structure. But in comparison with the perovskite family of structures, the symmetry of the paraelectric or prototype symmetry phase is low, which limits the number of domain variants in each crystal and limits the effectiveness of the poling process in ceramic materials. Furthermore, the low polarizability in many of these structures and systems also contributed to low piezoelectric coefficients. Examples of such structures include tungsten bronze, Aurivillius phases, perovskite layer structure (PLS) and lithium niobate. Another common finding in ferroelectric materials is the presence of lead or bismuth. Non-MPB materials exhibit T_C's of up to 1000°C, however the limited number of polar directions or limited polarizations lead to low piezoelectric d_{33} values.

Non-MPB perovskite piezoelectrics such as $BaTiO_3$(BT), $(Na_{1/2}Bi_{1/2})TiO_3$ (NBT), and modified-PT compositions have large piezoelectric (d) coefficients (for non-MPB materials) and low T_C's. Perovskite BT has limited use as a piezoelectric material, due to multiple ferroelectric-ferroelectric phase transitions, a low T_C of 120°C, and the resulting large temperature dependence of piezoelectric properties.

NBT based materials have recently gained interest due to environmental concerns regarding lead-containing materials. It has a T_C=335°C and relatively small piezoelectric d_{33} coefficient, large for a non-lead-containing piezoelectric. Tetragonal lead titanate, $PbTiO_3$ (PT), has a Tc of 490°C. In Lead titanate, reducing the tetragonality has the effect of reducing T_C. Some modified lead titanate materials maintain the advantage of a relatively large piezoelectric coefficients, improved temperature dependence and a large coercive field making it suitable for high power applications while offering improved resistance to high temperature aging.

The tungsten-bronze family materials, such as lead metaniobate ($PbNb_2O_6$), have been used in the manufacture of medical ultrasound and non-destructive testing (NDT) transducers, due to their high bandwidth (low QM). (Ba, Pb)Nb_2O_6 with about 10% Ba is a common composition with a T_C=400°C. However, low electrical resistivity, poor density, and low mechanical strength limit the widespread application of the tungsten-bronze family materials. Single crystal materials such as α-Quartz (SiO_2) and lithium niobate ($LiNbO_3$) are also widely used commercial piezoelectrics. Quartz is a non-ferroelectric piezoelectric crystal familiar for its use as a frequency standard. Lithium niobate is ferroelectric with a very high T_C=1150°C. The large piezoelectric voltage coefficient makes lithium niobate an ideal material for sensor applications. Usage temperature is generally limited to 650°C due to low electric resistivity and loss of oxygen at high temperatures.

Layer structured ferroelectrics consist of periodic stacks of perovskite-like and non-perovskite layers. Bismuth layer structure ferroelectrics (BLSF), also called the Arrivillius family (with the general formula An+1BnOn+3), such as $Bi_4Ti_3O_{12}$ have alternating layers of perovskite like units and bismuth oxide layers. $Na_{0.5}Bi_{4.5}Ti_4O_{15}$, another member of the BSLF family, has been reported with T_C>600°C. However, low symmetry and generally plate like morphology limit the ability to pole these materials resulting in low piezoelectric coefficients.

Perovskite layer structure ferroelectrics (PLS), with the general formula $A_2B_2O_7$, consist of perovskite like slabs containing corner sharing BO_6 octahedral and twelve-fold coordinated A-site cations, slabs are 4 octahedra thick and linked by A-site cation layers at the top and bottom slab boundaries. PLS ferroelectrics have very high Curie temperatures with T_C from 800 °C to 1200°C.

Therefore, a number of piezoelectric materials suitable for high temperature applications have been presented. The strength of the non-MPB materials is the availability of high T_C materials (T_C>500°C) for sensing applications. Sensing applications utilize the large g coefficients found in a number of these materials. Piezoelectric ceramics in the PLS and related families, in particular, have large voltage coefficients and high resistivity. However, the low piezoelectric strain coefficients limit actuator applications, which require large E-field induced strains over a broad temperature range.

REFERENCES

[1] David Michael Fanning, Structure Property Relations In Ferroelectric Materials, Thesis, University of Illinois at Urbana-Champaign, 2000.

[2] A J Moulson and J M Herbert, *Electroceramics: Materials, Properties, Applications*, (Chapman and Hall, London, 1990).

[3] N de Mathan, E Husson, G Calvarin, J R Gavarris, A W Hewat, A Morell, *J. Phys.: Condens. Matter* 3, 8159 (1991).

[4] V A Isupov, *Ferroelectrics* 90, 113 (1989).

[5] K Uchino, *Ferroelectrics* 151, 321 (1994).

[6] L E Cross, S J Jang, R E Newnham, S Nomura and K Uchino, *Ferroelectrics* 23, 178 (1990).

[7] Z-G Ye, *Ferroelectrics* 184, 193 (1996).

[8] G A Smolenskii, A I Agranovskaya, *Sov. Phys. -- Tech. Phys.*, 3 [7] 1380 (1958).

[9] L E Cross, *Ferroelectrics* 76, 241 (1987).

[10] D Viehland, S J Jang, L E Cross and M Wuttig, *J. Appl. Phys.* 68, 2916 (1990).

[11] G Fulcher, *J. Am. Cer. Soc.*, 8, 339 (1925).

[12] T. Granzow, *Phys. Rev. Lett.* 92 065701-1 (2004).

[13] G. Sumara, *J. Phys., Conds. Matter* 15 R367(2003).

[14] L.E. Cross, *Ferroelectrics*, 151, p305 (1994).

[15] L.E. Cross, *Ferroelectrics*, 76, p241(1987).

[16] V. Bobnar et. al. *Phys. Rev. Lett.* 84 5892 (2000).

[17] Park. S. E., Shrout. T.R., *J. Appl. Phys.*, 1997; 82: 1804.

[18] Dong. M., Ye. Z. G, *J. Crystal Growth*, 2000; 209: 81.

[19] David. R., Gabrielle. G., X-ray topography, NIST Recommended Practice Guide, Special publication 960-10.

[20] Huang. X, Jiang. S, Zeng. W, Hu. X, Feng. D, *Appl. Phys. Lett.*, 1995; 66:2649.

[21] Zhao. J, Yang. P, Jiang. S, *Appl. Phys. Lett.*, 1991; 59:1952.

[22] Cloetens. P, Boller. E, Ludwig. W, Baruchel. J, *europhysics news*, 2010; March:46.

[23] Jiang. J, Zhao. J, Tian. Y, *Instr. and Meth.*, 1993; 366: 354.

[24] Xiao. J, Zhang. X, Zhu.P, Huang. W, Yuan. Q, *Solid State Communication*, 2008; 148: 109.
[25] Yuan. Q, Zhao. C, Luo. W, Yin. X, *Journal of Crystal Growth*, 2001; 233:717.
[26] Fujishiro, K.; Vlokh, R.; Kiat, J.; Dkhil, B.; Yamashita, Y. *Jpn. J. Appl. Phys.* 1998, *37*, 5246-5248.
[27] Xiao, J.; Tian, Y.; Huang, W.; Hang, Y.; Yin, S. *Phys. Lett. A.* 2002, *300*, 456-460.
[28] Paszkowicz, W. *Nuclear Instruments and Method in Physics Research B.*2002, *198*, 142-182.
[29] Kreisel, J.; Bouvier, P.; Naglione, M.; Dkhil, S. A. *Phys. Rev. B.* 2004, *69*, 090104.
[30] Kreisel, J.; Dkhil, B.; Bouvier, P.; Kiat, *J. Phys. Rev. B.* 2002, *65*, 172101.
[31] Janolin, P. E.; Dkhil, B.; Bouvier, P.; Kreisel, J.; Thomas, P. A. *Phys. Rev. B.* 2006, *73*, 094128.
[32] Liu, J.; Li, Y. *Phys.: Condens. Matter.* 2002, *14*, 10505-10509.
[33] Tinite, S.; Rabe, K.; Vanderbilt, D. *Phys. Rev. B.* 2003, *68*, 144105.
[34] Bellaiche, L.; Kunc, K.; Besson, J. *Phys. Re. B.* 1996, *54*, 8945-8959.
[35] Zhao, J.; Ross, N.; Angel, R. *Acta Crystallographica Section B.* 2004, *60*, 263-271.
[36] Hazen, R. *Rev. Mineral.* 1985, *14*, 317-346.
[37] Kreisel, J.; Glazer, A. *J. Phys.: Condens. Matter.* 2000, *12*, 9689-9698.
[38] V.V.S.S. Sai Sunder, and A. M. Umarji, *Material Research Bulletin*, 30, 427-434, (1995).
[39] S. P. Singh, A. K. Singh, and D. Pandey, *J. Phys.: Condense Matter.* 19, 036217, (2007).
[40] S. P. Singh, A. K. Singh, and D. Pandey, *Ferroelectrics*, 324, 49 (2005).
[41] C.-S. Tu, C. T. Tseng, R. R. Chien, V. Hugo Schmidt, and C.-M. Hsieh, *J. Appl. Phys.* 104, 054106 (2008).
[42] Jie Wang, X. G. Tang, H. L. W. Chan, C. L. Choy, and Haosu Luo, *Appl. Phys. Lett.* 86, 152907, (2005).
[43] Kui Liu, Xinyi Zhang, Shiqiang Wei, Jingzhong Xiao, Dielectric anomalies induced by different mechanisms in Pb(Fe1/2Nb1/2)1-xTixO3 single crystals, arXiv: 0810.1346(October 2008).
[44] Emiel Hensen, Qingjun Zhu, Pang-Hung Liu, Kuei-Jung Chao, and Rutger van Santen, *Journal of Catalysis* 226, 466–470 (2004).
[45] Richard E. Eitel, "Novel piezoelectric ceramics:development of high temperature, high performance piezoelectric on the basis of structure", Thesis in Department of Materials Science and Engineering, The Pennsylvania State University, December 2003.
[46] H. Lammer, "Flexible Piezoelectric Films, " International Patent No. WO 02081401 (October 17 2002).
[47] S. Yoshikawa, A. Bogue and B. Degon, "Commercial Application of Passive and Active piezoelectric Vibration Control, " in Proceedings of the *11th IEEE International Symposium on Applications of Ferroelectrics*, (eds. E. Colla, D. Damjanovic and N. Setter) Montreux, Switzerland, 293-4 (1998).
[48] B. Jaffe, W. R. Cook and H. Jaffe, *Piezoelectric Ceramics,* Academic Press, New York, (1971).
[49] D. A. Berlincourt, C. Molik and H. Jaffe, "Piezoelectric Properties of Polycrystalline Lead Titanate Zirconate Compositions, " *Proceedings of the IRE* [2] 220-9 (1960).
[50] S.-E. Park, T.R. Shrout, *J. Appl. Phys*. 82 (1997) 1804.

[51] M. Dong, Z.-G. Ye, *J. Crystal Growth* 209 (2000) 81.
[52] N. Yasuda, H. Ohwa, M. Kume, Y. Hosono, Y. Yamashita, S. Ishino, H. Terauchi, M. Iwata, Y. Ishibashi, *Jpn. J. Appl. Phys*. 40 (2001) 5664.
[53] H.S. Luo, G.S. Xu, Z.W. Yin, *Jpn. J. Appl. Phys*. 39 (2000) 5581.
[54] S. J. Zhang, D. Y. Jeong, Q. M. Zhang, and T. R. Shrout, *J. Cryst. Growth* 247, 131 (2003).
[55] S. J. Zhang, L. Lebrun, S. Rhee, R. Eitel, C. A. Randall, and T. R. Shrout, *J. Cryst. Growth* 236, 210 (2002).
[56] S. J. Zhang, C. A. Randall, and T. R. Shrout, *Jpn. J. Appl. Phys*., Part 2 42, L1152 (2003)
[57] S. J. Zhang, C. A. Randall, and T. R. Shrout, *Appl. Phys. Lett*. 83, 3150 (2003).
[58] R. E. Eitel, C. A. Randall, T. R. Shrout, P. W. Rehrig, W. Hackenberger, and S.-E. Park, *Jpn. J. Appl. Phys*., Part 1 40, 5999 [2001].
[59] R. E. Eitel, C. A. Randall, T. R. Shrout, and S.-E. Park, *Jpn. J. Appl. Phys*., Part 1 41, 2099 (2002).
[60] Y. Shimojo, R. Wang, T. Sekiya, T. Nakamura, and L. E. Cross, *Ferroelectrics* 284, 121 (2003).
[61] Yoshiyuki Inaguma, et al, "High-pressure synthesis and ferroelectric properties in perovskite-typeBiScO3–PbTiO3 solid solution", *J. Applied Physics*, 95, 231 (2004).
[62] R. E. Eitel, C. A. Randall, T. R. Shrout, P. W. Rehrig, W. Hackenberger and S.-E. Park, *Japanese Journal of Applied Physics*, 40, 5999 (2001).
[63] Takeshi Yoshimura, *Appl. Phys. Lett*., 81 (2002) 2065.
[64] J. Nino and S. Trolier-McKinstry, *Journal of Materials Research*, 19, 568 (2004).
[65] H. Wen, X. Wang, X. De ng, L. Li, *Appl. Phys. Lett*., 88, 222904 (2006).
[66] H. Wen, X. Wang, X. De ng, t. Sun, L. Li, *J. Appl. Phys*., 101, 016103(2007).
[67] H. Wen, X. Wang, L. Li, *J. Am. Ceram. Soc*. 89, 2345 (2006).
[68] H. Wen, X. Wang, L. Li, *J. Appl. Phys*., 100, 084315 (2006).
[69] H. Wen, X. Wang, C. Zhong, L. Shu, L. Li, *Appl. Phys. Lett*., 90, 202902(2007).
[70] J. Scott, C. A. Araujo, *Science* 246, 1400 (1989).
[71] Y. Li, W. Chen, Q. Xu, J. Zhou, X. Gu, *Materials Letters*, 59, 1361 (2005).
[72] T. Oikawa, M. Aratani, H. Funakuba, K. Saito, M. Mizuhira, *J. Appl. Phys*., 95, 3111 (2004).
[73] Y. Inauguma, A. Miyaguchi, M. Yoshida, T. Katsumata, *J. Appl. Phys*., 95, 231 (2004).
[74] J. Zeng, M. Zhang, L. Wang, C. Lin, *J. Phys.: Condens. Mater*., 11, 1139 (1999).

Chapter 3

SIMPLE INTRODUCTION FOR HIGH-TC FERROELECTRIC CERAMICS

Lan Xu and Jingzhong Xiao
Physics Department, University of Coimbra, Portugal

The research of Bi-based piezoelectric ceramics are very important because of their excellent properties, especially their high Curie temperature for use at high temperatures. To get a higher Curie temperature and more excellent properties, the various properties of BSPT and doped with element or compound in Bi-based piezoelectric materials were summarized in this paper.

The ferroelectric $Pb(Zn_{1/3}Nb_{2/3})O_3$-$PbTiO_3$(PZNT) and $Pb(Mg_{1/3}Nb_{2/3})O_3$-$PbTiO_3$ (PMNT) crystals, [PaSh97] have attracted extraordinarily interest for their ultrahigh piezoelectric constant (d_{33}>2000pC/N) and electromechanical coupling factors (k_{33}>94%), which makes them very promising in the electronic industries. However, one evident drawback is the low Curie temperature (for PZNT and PMNT, T_c~140-170°C), which limits their utilization in a higher temperature environment such as the automotive and aerospace industries, for the loss of ferro/piezo-electric response through depoling across T_c. Since the high Curie temperature and the temperature stability of the electrical properties as well as piezoelectric activity are very important, discovery and development of new high-T_c piezoceramic systems are especially attractive. Recently, bismuth-based perovskite systems $Bi(Me)O_3$–$PbTiO_3$ (*Me*= Me′ Me″=Sc^{3+}, In^{3+}, Yb^{3+} etc.) were found to possess much higher Curie temperatures in keeping with the end-member tolerance factor concept. [1-3] The Me′ and Me″ are octahedrally coordinated transition-metal cations that can be stoichiometrically combined to give an average B-site cation +3 valence. There are a large number of binary systems that can be suggested, as partially listed in [3].

This part of the chapter on piezoelectric materials introduces the recent progress of the $BiMeO_3$-$PbTiO_3$ ceramics. The present situation and future outlook for these materials are also presented.

The Development of $BiMeO_3$-$PbTiO_3$

To get a higher Curie temperature and more excellent properties, the various properties of BSPT and doped with element or compound in Bi-based piezoelectric materials were studied.

The Development of $BiScO_3$-$PbTiO_3$

The BSPT phase diagram, the structure of BSPT and the properties about application of BSPT were reported. [1-4]

BSPT Phase Diagram

To further study the properties of the BSPT, in 1969, Y. Shimojo et al studied the MPB phase diagram and the ferroelectric properties in the $BiScO_3$-$PbTiO_3$ system and the BS-PT system has MPB between 35 and 37.5%BS. [4] Due to the lack of stabilization of a perovskite-type solid solution (1-x)$PbTiO_3$-x$BiScO_3$ with $x \geq 0.45$, Yoshiro Shimojo et al synthesized it under high-pressure and further completed the phase diagram in 2004. It was shown that the crystal symmetry of the perovskite subcell changes in turn from tetragonal, to rhombohedral, to pseudocubic, to monoclinic, and then to triclinic as x increases and the tetragonal, rhombohedral, and pseudocubic phases are ferroelectric, while the monoclinic phase is not. In the morphotropic phase boundary it is in the vicinity of x=0.37 between tetragonal and rhombohedral phases. In addition, relaxor behavior in the dielectric constant was observed in the vicinity of x=0.5. The phase diagram and the variation of lattice parameters and peroskite parameters (a_p) versus x are presented. [5] The a_p denotes the cube root of volume of the perovskite subcell. In their study, the existence of intermediate phases between tetragonal and rhombohedral phases could not be detected at room temperature.

In the same year, the phase diagram for the perovskite BS-PT system was revisited using variable-temperature electron-diffraction and dielectric measurements on single crystals. R. E. Eitel et al contributed to the reports on revisions to the (1−x)$BiScO_3$-x$PbTiO_3$ phase diagram specifically, (i) the compositional dependence of the octahedral tilt-transition temperature (x<0.62 mol fraction PT), as determined using variable-temperature transmission electron microscopy, (ii) the high temperature curvature of the rhombohedral-tetragonal morphotropic phase boundary determined using dielectric measurements on single crystals, and (iii) Curie temperatures in the tetragonal phase field, which exceed that of the $PbTiO_3$ end-member. The phase diagram for the perovskite (1-x)$BiScO_3$-x$PbTiO_3$ system representing revisions based on the TEM and single-crystal data had been reported. In 2007, a structural study of the morphotropic phase boundary system (1−x)$BiScO_3$−x$PbTiO_3$ in which both A and B sites of the perovskite are randomly occupied was studied. A M_A monoclinic phase coexisting with a tetragonal phase has been evidenced for concentration 0.60<x<0.70. No evidence of local cations ordering nor chemical inhomogeneity has been found. Taking into account data of Eitel et al. [Jpn. J. Appl. Phys., Part 1 40, 5999(2001); Jpn. J. Appl. Phys. 41, 2099 (2002)] and our results, a global phase diagram has been drawn.

The Structure of BSPT

Since 2003, transmission electron microscopy has been used to study changes in domain structure and symmetry as a function of x by Randall et al. $BiScO_3$-$PbTiO_3$ has typical FE domain structures in the R and T phases. In the R phase, 71°and 109°domains were found to have twin planes on {100} and {110} planes, respectively. Zig-zag walls were also found consistent with earlier studies. T domain structures were observed to have 90° and 180° walls. In 2007, Chol-Jun Yu et al. [6] designed averaging of the pseudo-wavefunctions. [6] As for physical and numerical considerations, there is a subtle arbitrariness in constructing the Kleinman–Bylander pseudopotential. They construct the orthonormalized pseudo-wavefunctions as well as pseudopotentials of the virtual atom by averaging and providing reasonable properties for heterovalent solid solutions which has the advantage of great simplicity. Their results for the perovskite ferroelectrics (1− x)$BiScO_3$–x$PbTiO_3$ satisfy Vegard's law well as regards the cubic lattice constants and display improved values as regards the structural and electrical properties compared with previous work.

In 2010, a full structural study of the solid solution x$BiScO_3$-(1−x)$PbTiO_3$ (BS-PT) was been carried out by neutron powder diffraction using the General Materials Diffractometer (GEM) at the ISIS facility in Oxford and complemented by in-house x-ray diffraction studies and high temperature dielectric measurements. [7] In the phases of the BS-PT system as a function of composition, three different regions were identified where the high temperature cubic phase stabilizes into (1) a tetragonal phase for $x\leq 0.35$; (2) a tetragonal+monclinic phase for $0.36\leq x\leq 0.38$, and finally (3) to an apparent monoclinic phase for $x\geq 0.39$. The suggestion of monoclinic symmetry for composition $x\geq 0.39$ has been further supported by the high temperature x-ray diffraction results. In addition, a possible correlation has been shown between the enhancement of T_c and the shift of the Ti/Sc atom from its ideal position.

Piezoelectric devices are being developed for extreme-temperature applications, including small-signal sensors, high-signal acoustic transducers, and actuators for temperatures higher than 400 °C. These devices may find applications in space exploration, electric aircraft, oil and geothermal well drilling tools, and automotive smart brakes. In practice, all applications of piezoelectric materials limit usage temperature to $T<\frac{1}{2}T_C$, due to the loss of polarization, typically at temperatures less than 200°C. [8]

Various Properties of BSPT Were Studied as Follow

The poled piezoelectric ceramics often begin to lose their piezoelectric properties at a temperature somewhat below the Curie temperature due to the thermal instability of polarization. Stability of polarization at the temperature is very important. BS-PT with a tetragonal phase showed good resistance to thermal depoling up to 440°C, close to their Curie temperature. [9]

Dielectric spectra of high temperature piezoelectric (1-x)$BiScO_3$- x$PbTiO_3$ of composition near the MPB ($x\cong 0.64$) were investigated in the frequency range of 100 Hz–1THz at temperatures between 10 and 900 K. [10] The dielectric response in a broad frequency range is not only near T_c but also below room temperature. BSPT x=0.62–0.66

ceramics (density≥96%) and x≌0.62 single crystal preparation can be found in [11] and [12] (x=0.62, 0.64, 0.66).

The nonlinear dielectric and converse piezoelectric response in (1−x)$BiScO_3$–x$PbTiO_3$ (BS-PT) ferroelectric ceramics were measured and analyzed by R. E. Eitel et al [13]. A ferroelectric phase dependence on the magnitude of the nonlinear was contributed to the dielectric and piezoelectric responses. The magnitudes of the nonlinear coefficients, which quantify the irreversible extrinsic response, were found to be smallest in the ferroelectric tetragonal phase, larger in the rhombohedral composition, and highest at the MPB. Their work provides strong evidence that the same loss mechanism dominates the nonlinear portion of both the dielectric and piezoelectric responses in piezoelectric ceramics.

In 2006, homogeneous and nanosized (1-x)$BiScO_3$-x$PbTiO_3$ ceramic powders (10 nm) synthesized by the sol-gel method [14-16], were obtained by calcining the dried gel at 600°C for 3 h, and the nano-powders can be sintered at 1000°C, which is 100°C lower than those prepared by the traditional method. Using nanosized powder, ultrafine grained BSPT ceramics with an enhanced piezoelectric coefficient were achieved.

In 2009, M. Algueró et al [17] studied the macroscopic electrical properties of highly dense, nanostructured ceramics of $BiScO_3$–$PbTiO_3$ with high Curie temperature and piezoelectric activity. The nanostructured material still presents the ferroelectric transition above 700K. Macroscopic ferroelectric switching is clearly demonstrated in a nanostructured ceramic, with figures of remnant polarization as high as 19μC cm^{-2}. Ceramic disks of the nanostructured material can be poled, and a d_{33} piezoelectric coefficient of 40 pC/ N resulted. Their radial piezoelectric resonance was also excited and analyzed, and a d_{31} coefficient of −14 pC/ N and a coupling factor of 9% were obtained.

To get higher ferroelectric properties of (1−x)$BiScO_3$–x$PbTiO_3$ ceramics, Tingting Zou et al [18] utilized two-step sintering. Their ferroelectric properties were as follows. The existence of a morphotropic phase boundary at the composition x=0.635, which exhibits a piezoelectric coefficient d_{33} of 700 pC/N at room temperature. Furthermore, a higher electromechanical coupling factor K_p=0.632 and a larger remnant polarization P_r=47.3μC/cm^2 were obtained. The paraelectric-to-ferroelectric phase transition occurs at 446°C, slight lower than in the coarse grain ceramics with a similar composition, suggesting a grain size effect. The local effective piezoelectric coefficient d_{33}^* was estimated to be 795pC/N at 2.29V, measured by scanning probe microscopy.

Scaling behaviors of the dynamic hysteresis of 40$BiScO_3$–60$PbTiO_3$ bulk ceramic were studied by Gang yu et al [19] in 2010. The scaling relations of the hysteresis loop area <A> against frequency f and amplitude E_0 for the saturated loops take the form $\langle A\rangle \propto f^{0.05}E_0^{0.6}$ when the frequency goes to zero, and the form turns into $\langle A\rangle \propto f^{-0.01}E_0^{0.3}$ as the frequency f tends to infinity. Combining the information from these two relationships, they draw the conclusion that the area <A> first grows and then decays with increasing frequency f over a wide range.

In 2010, the structure of $PbTiO_3$-$BiScO_3$ in the tetragonal phase was refined by means of neutron powder diffraction (NPD) by Jun Chen et al [20]. High-T_C piezoelectric (1−x)$PbTiO_3$-x$BiScO_3$ shows a nonmonotonic trend of T_C in the tetragonal phase with respect to content of $BiScO_3$. The cation displacement of Pb/Bistays at a high level, while that of Ti/Sc distributes in a wider range. The nonmonotonic trend of T_C is quantitatively correlated to polarization. The general character of high-T_C in PT–$BiMeO_3$ results from the unique role of Bi substitution.

The Development of BS-PT-Relaxor

As we known, The perovskite (1−x)$BiScO_3$-x$PbTiO_3$ (BS-PT) system exhibits a large piezoelectric coefficient (d_{33}>460pC/N in polycrystalline form) and a high Curie temperature T_C=450°C in the vicinity of the morphotropic phase boundary (x=0.64mol fraction $PbTiO_3$), which separates the rhombohedral and tetragonal phases. [21] To improve and enhance the ferroelectric properties, the adulteration of the element or compound is performed.

In 2002, because of the low mechanical quality factor (Q_m) of BSPT, which may be a serious problem for resonance mode applications, Junghu Ryu et al [22] developed the (0.9-x)BiScO3-xPbTiO3-0.1$Pb(Mn_{1/3}Nb_{2/3})O_3$ (BS-PT- PMnN) with a high Curie temperature and high Q_m piezoelectric. A morphotrophic phase boundary (MPB), separating the rhombohedral and tetragonal phase was found at x~0.6, where the highest piezoelectric constant (d_{33}) and electromechanical coupling factor (k_p) were obtained. A high mechanical quality factor (Q_m) which is important for high power applications, and high transition temperature (T_c) were obtained from the tetragonal phase (x~0.68), and the magnitude of these constants were 1000, and 420°C, respectively. The maximum vibration velocity 0.72 m/s was obtained from x=0.56 composition. Compared to the unmodified BS–PT ceramic there is significant drop in the magnitude of k_p and d_{33}. [23] The magnitude of Q_m increased significantly as compared to unmodified BS–PT single crystal in the tetragonal region. [24] The maximum Q_m was obtained at 68% PT and the value was around 1000. With this composition the magnitude of d_{33} and k_p are ~100 pC/N and 0.22 respectively. A high mechanical quality factor, high transition temperature (T_{max}), and high vibration velocity (υ_0) were obtained from this solid solution system. The magnitude of these characteristics were 1013, 420°C, and 0.72 m/s, respectively.

In 2003, Junghu Ryu et al [25] studied 0.3$BiScO_3$–0.6$PbTiO_3$–0.1$Pb(Mn_{1/3}Nb_{2/3})O_3$ ferroelectric material with a high dielectric maximum of ~350°C. The dielectric constant in the low-temperature region of 10°C<T<150°C is found to be independent of the frequency with a temperature gradient of $\partial\varepsilon/\partial T$=2.5/°C. The room-temperature magnitude of dielectric constant is ~600.

In 2005, Sterianou et al [26] studied the (1−x)$BiSc_{1/2}Fe_{1/2}O_3$-x$PbTiO_3$ (BSF-PT) ceramics in which a MPB occurred between rhombohedral and tetragonal phases at x≈0.50 (T_C=440°C). Samples close to the MPB gave piezoelectric and electromechanical coupling coefficients of ~300 pC/N and 0.5, respectively.

In 2005, Soon Mok Choi et al [27] studied the binary solid-solution system (1-x)$Bi(Ni_{1/2}Ti_{1/2})O_3$–x$PbTiO_3$ [(1−x) BNT–xPT] for 0.39≤x≤1.00. MPB was found at the composition 0.51BNT-0.49PT, with a corresponding paraelectric-ferroelectric phase transition T_C≈400°C and piezoelectric d_{33} coefficients≈260 pC/N at room temperature. MnO_2 doping to the BNT-PT system close to the MPB compositions, with an improvement in dielectric losses but a reduction in d_{33}≈180pC/N and a T_C enhancement above T_C~495°C was also studied.

In 2005, bismuth-based perovskite solid solution BNN–PT has been studied by Zhang et al [28], a MPB was found to be located in the region of 62.5%–65% PT content, with a T_c

value around 260–273°C. The highest room-temperature dielectric constant was found to be 1380 for the BNN–PT62.5 material, while the highest piezoelectric coefficient and electromechanical coupling factor were found to be 140 pC/N and 22.4%, respectively, for the BNN-PT65 ceramic. Improved dielectric loss(~1.2%) and Q value(~140) were obtained for the Mn-modified BNN-PT65 material due to the hardening effect. The addition of manganese(Mn) resulted in lowering the dielectric loss and increasing the mechanical quality factor Q when compared with the pure counterpart. The coercive field was found to increase to 30.7 kV/cm when BNN-PT65 was modified with magnesium (Mg).

Effects of Fe substituents on bismuth scandium lead titanate ceramics (1-x)BiScO3-xPbTiO3(BSPT) at x=0.64 were investigated by Winotai et al [30] in order to improve the piezoelectric properties. The Fe-doped BSPT piezoelectrics clearly show “soft” characteristics for their smaller P-E hysteresis loops with increasing Fe content and lower electric field coercivity E_c. The data exhibit the increase in the mechanical quality factor (Q_m) while the dielectric constant (K_{33}^T) decreases compared with those of the undoped sample. The maximum of Q_m is reached at 526.41 for the 1.00mol% Fe-doped BSPT. Moreover, the dopant slightly increases the piezoelectric constant (d_{33}) and planar coupling coefficient (k_p). The Curie temperature (T_c) increases with Fe content and the maximum T_c value reaches 451°C for 2.00mol% Fe doping in comparison to 437 ◦C of the undoped BSPT ceramics.

In 2005, the shear mode for Mn-modified BSPT ceramics near the MPB was studied by Shujun Zhang et al [31]. The resistivity and time constant were $3\times10^7\Omega$cm and 0.08s, respectively, at 450°C for modified BSPT66. The dielectric constant K_{11}^T and dielectric loss were found to be 1112 and 1%, respectively, at room temperature, showing a Curie temperature at 468°C. The electromenchanical coupling factor k_{15} was calculated to be 61%, staying nearly constant up to 440°C. The detailed properties of manganese modified BSPT materials are summarized and compared to PZT5A.

In 2007, compositions near MPB of the $BiScO_3$–$Pb(Mg1/3Nb_{2/3})O_3$–$PbTiO_3$ ternary system with high temperature relaxor properties (1 kHz) of T_{max} ~250°C and a permittivity maximum of ~16000 were studied by Craig J. Stringer and Clive A. Randall [31].

In 2007, the structure and dielectric properties of (1-x)$Bi(In_{0.20}Ga_{0.05}Sc_{0.75})O_3$-x$PbTiO_3$ (x = 0.55~0.70) ceramics (BIGS-PT) were investigated by Chen et al [32]. The MPB appears in BIGS-PT for x = 0.60 and a high Curie temperature is 106°C. BIGS-PT ceramics show good piezoelectric and ferroelectric properties: d_{33} = 295 pC/N, k_P = 43.0%, Pr = 19.2 *C/cm*^{2}, and E_c = 34.9 *kV/mm*, respectively.

In 2007, $Bi(Zn_{1/2}Ti_{1/2})O_3$-$BiScO_3$-$BaTiO_3$,which is a very stable dielectric constant greater than 1000 and low loss tangents(<0.01) over a wide range of temperature (T<400 °C) were studied by Chien-chih Huang et al [33].

In 2008, Jiang et al [34] studied $BiScO_3$–$LiSbO_3$–$PbTiO_3$ solid solutions. In the vicinity of the morphotropic phase boundary, the ceramics showed enhanced piezoelectric and ferroelectric properties: piezoelectric constant d_{33}=545 pC/N, planar electromechanical coupling factors k_p =58%, remnant polarization P_r=28.3μC/cm^2, and the coercive field E_c=1.8kV/mm, respectively. The temperature of the permittivity maximum (T_{max}) increased with an increasing measure of frequency, over the range of 300-340°C.

In 2008, cobalt-doped (1−x)$BiScO_3$-x$PbTiO_3$ piezoelectric ceramics were studied by Zhonghua Yao et al [35], in which a MPB region was observed in the compositional range of 0.48≤x≤0.53 from the phase structure measurement. The ferroelectric phase transition of the composition x=0.50 near the MPB was found to be a displacive and first-order transition with excellent piezoelectric properties, T_m=470 °C, d_{33}=282pC/N, K_p=0.49. Dielectric and piezoelectric properties of (1-x)$Bi(Sc_{0.5}Co_{0.5})O_3$-x$PbTiO_3$ ceramics for composition near the MPB compared to undoped PZT and other BiScO3-based ceramics were investigated.

In 2009, 0.36$BiScO_3$–0.64{(1−x)$PbTiO_3$–x$LiNbO_3$} (BSPTLNx) ceramics with perovskite structure were developed by Yi Chen et al [36-39]. The maximum electromechanical values of d_{33} =465pC/N and k_p =0.57 were gained in BSPTLNx for x=0.04, as well as the highest remnant polarization P_r =48μC/cm^2. However, the T_c gradually decreases with increasing LN content in the BSPTLNx system.

In 2009, Hideki Ogihara et al [40] investigated the (1-x)$BaTiO_3$-x$BiScO_3$ (x=0-0.5) ceramics to acquire a better understanding of the binary system, including determination of the symmetry of the phases, the associated dielectric properties, and the differences in the roles of Bi_2O_3 and $BiScO_3$ substitutions in a $BaTiO_3$ solid solution. A systematic structural change from the ferroelectric tetragonal phase to a pseudo-cubic one was about x=0.05–0.075 at room temperature. a gradual change from proper ferroelectric behavior in pure $BaTiO_3$ to highly diffusive and dispersive relaxor-like characteristics from 10 to 40 mol% $BiScO_3$. Quantification of the relaxation behavior was obtained through the Vogel–Fulcher model, which yielded an activation energy of 0.2–0.3eV. The attempt characteristic frequency was 10^{13} Hz and the freezing temperature, T_f, ranged from -177° to -93 ° C as a function of composition. Several of the compositions showed high relative permittivities with low-temperature coefficients of capacitance over a wide range of temperature. The high coercive fields, low remanent polarization, and high activation energies suggest that in the $BiScO_3$–$BaTiO_3$ solid solutions, the polarization in nanopolar regions is weakly coupled from region to region, limiting the ability to obtain long-range dipole ordering in these relaxors under field-cooled conditions.

CONCLUSION

The properties of various perovskite Bi-based piezoelectric ceramics were investigated. The Bi bismuth-based perovskite systems Bi(*Me*)O_3–$PbTiO_3$ (*Me*= Me′ Me″=Sc^{3+}, In^{3+}, Yb^{3+} etc.) were found to possess much higher Curie temperatures. There was a focus on the $BiScO_3$-$PbTiO_3$ with high T_c. The BSPT phase diagram, the structure, the various properties and the application were shown in this work. To enhance the piezoelectric properties, the Bi(*Me*)O_3–$PbTiO_3$ doped with element and compound were also investigated. These Bi-based piezoelectric ceramics may find applications in space exploration, electric aircraft, oil and geothermal well drilling tools, and automotive smart brakes.

REFERENCES

[1] Y. Shimojo, R. Wang, T. Sekiya, T. Nakamura, and L.E. Cross. *Materials Research*, 1297, p.28(1969).

[2] Yoshiro Shimojo, Atsushi Miyaguchi, Masashi Yoshida, Tetsuhiro Katsumata , Yoshiro Shimojo, Ruiping Wang, and Tadashi Sekiya, *Journal of Applied Physics*, 95 (2004)231.

[3] R. E. Eitel, S. J. Zhang, T. R. Shrout, C. A. Randall and I. Levin, *Journal of Applied Physics*, 96, (2004), 2828.

[4] J. Chaigneau, J. M. Kiat, C. Malibert, and C. Bogicevic. *Physical Review* B 76, 094111 (2007).

[5] C. A. Randall, R. E. Eitel, T. R. Shrout, D. I. Woodward, I. M. Reaney, *Journal of Applied Physics*, 93 (2003)9271.

[6] Chol-Jun Yu and Heike Emmerich. *J. Phys.: Condens. Matter* 19 (2007) 306203 (13pp).

[7] ´I˜ niguez J, Vanderbilt D and Bellaiche L 2003 *Phys. Rev*. B 67 224107.

[8] S´aghi-Szab´o G, Cohen R E and Krakauer H 1998 *Phys. Rev. Lett*. 80 4321.

[9] K. Datta, D. Walker, and P. A. Thomas. *Physical Review* B 82, 144108 (2010).

[10] T. R. Shrout, R. E. Eitel, and C. A. Randall, *Piezoelectric Materials in Devices*, edited by N. Setter (EPFL Swiss Federal Institute of Technology, Switzerland, 2002), p. 413.

[11] Si Chen, Xianlin Dong, Chaoliang Mao, and Fei Cao. *Journal of the American Ceramic Society*, 89(2006)3270.

[12] V. Porokhonskyy, S. Kamba, A. Pashkin, M. Savinov, J. Petzelt, R. E. Eitel, C. A. Randall, *Applied Physics Letters*, 83(2003)1605.

[13] R. E. Eitel, C. A. Randall, T. R. Shrout, and S. E. Park, *Jpn. J. Appl. Phys.*, Part 1 41, 2099(2002).

[14] S. Zhang, L. Lebrun, S. Rhee, R. E. Eitel, C. A. Randall, and T. R. Shrout, *J. Cryst. Growth* 236, 210 (2002).

[15] R. E. Eitel, T. R. Shrout, and C. A. Randall, *Journal Of Applied Physics* 99, 124110 (2006).

[16] Wei Zhao, Xiaohui Wang, w Junjie Hao, Hai Wen, and Longtu Li. *Journal of the American Ceramic Society*. 89 (2006)1200.

[17] M. Algueró, H. Amorín, T. Hungría, J. Galy, and A. Castro. *Applied Physics Letters* 94, 012902 (2009).

[18] Tingting Zou, Xiaohui Wang, Han Wang, Caifu Zhong, Longtu Li, and I-Wei Chen. *Applied Physics Letters* 93, 192913(2008).

[19] Gang Yu, Xuefeng Chen, Fei Cao, Genshui Wang, Xianlin Dong. *Solid State Communications* 150 (2010) 1045–1047.

[20] Jun Chen, Krishna Nittala, Jacob L. Jones, Penghao Hu, and Xianran Xing. *Applied Physics Letters* 96, 252908(2010).

[21] R. E. Eitel, S. J. Zhang, T. R. Shrout, C. A. Randall and I. Levin, *Journal of Applied Physics*, 96, (2004), 2828.

[22] Jungho Ryu, Shashank Priya, Chiharu Sakaki and Kenji Uchino. *Japanese Journal of applied Physics*. 41 (2002)6040.

[23] R. E. Eitel, C. A. Randall, T. R. Shrout, P. Rehrig, W. Hackenberger and S.-E. Park: *Jpn. J. Appl. Phys*. 40 (2001) 5999.

[24] S. Zhang, L. Lebrun, S. Rhee, R. E. Eitel, C. A. Randall and T. R. Shrout: *J. Cryst. Growth* 234 (2002) 210.

[25] Jungho Ryu, Shashank Priya and Kenji Uchino, *Applied Physics Letters*, 82(2003)251.

[26] I. Sterianou, I. M. Reaney, D. C. Sinclair, D. I. Woodward, D. A. Hall, A. J. Bell and T. P. Comyn. *Applied Physics Letters*, 87(2005) 242901.
[27] Soon Mok Choi, Craig J. Stringer, Thomas R. Shrout, and Clive A. Randall. *Journal of Applied Physics*. 98(2005)034108.
[28] Shujun Zhang, Craig Stringer, Ru Xia, Soon-Mok Choi, Clive A. Randall, and Thomas R. Shrout. *Journal of Applied Physics*. 98, 034103 (2005).
[29] P. Winotai, Nitinai Udomkan, and Siwaporn Meejoo. *Sensors and Actuators* A. 122 (2005). 257–263.
[30] Shujun Zhang, Richard E. Eitel, Clive A. Randall, Thomas R. Shrout and Edward F. Alberta. *Appl. Phys. Lett.* 86, 262904(2005).
[31] Craig J. Stringer and Clive A. Randall. *J. Am. Ceram. Soc.*, 90 [6] 1802–1807 (2007).
[32] Yi Chen; Bao-Quan Qin ; Yi-Hang Jiang ; Zhuo Xu ; Qiang Chen ; Xi Yue; Ding-Quan Xiao; Jian-Guo Zhu. *Ferroelectrics*, 356(2007)40.
[33] Chien-Chih Huang and David P. Cann , Xiaoli Tan, Naratip Vittayakorn. Journal of Applied Physics 102, 044103 (2007).
[34] Yihang Jiang, Yi Zhao, Baoquan Qin, Yuzhi Jiang, Wei Shi, Lihua Li, Dingquan Xiao, and Jianguo Zhu. *Applied Physics Letters* 93, 022904(2008).
[35] Zhonghua Yao, Hanxing Liu, Yang Liu, Zhaohui Wu, Minghe Cao, and Hua Hao. *Applied Physics Letters* 92, 142905 (2008)
[36] R. E. Eitel, C. A. Randall, T. R. Shrout, and S. E. Park, *Jpn. J. Appl. Phys.,* Part 1 41, 2099(2002).
[37] J. R. Cheng, R. Eitel, N. Li, and L. E. Cross, *J. Appl. Phys*. 94, 605(2003).
[38] I. Sterianou, I. M. Reaney, D. C. Sinclair, D. I. Woodward, D. A. Hall, A. J. Bell, and T. P. Comyn, *Appl. Phys. Lett.* 87, 242901(2005).
[39] B. Jaffe, W. R. Cook, and H. Jaffe, *Piezoelectric Ceramics* (Academic, New York, 1971).
[40] Yi Chen, Jianguo Zhu, Dingquan Xiao, Baoquan Qin, Yihang Jiang. *Journal of Alloys and Compounds* 470 (2009) 420–423.
[41] Hideki Ogihara, Clive A. Randall, and Susan Trolier-McKinstry. *J. Am. Ceram. Soc.*, 92 [1] 110–118 (2009).

Chapter 4

HIGH-TC FERROELECTRIC SINGLE CRYSTALS

Since the development of perovskite ferroelectrics $BaTiO_3$ and $Pb(Zr_{1-x}Ti_x)O_3$ (PZT) in the 1950s [1], piezoelectrics have been the subject of numerous investigations, which are widely used in underwater sonar, modern medical ultrasound imaging and other high performance actuator and transducers. [1-12] For high performance piezoelectrics, the high Curie temperature ferroelectrics with the high piezoelectric activity are desired. Relaxor-based single crystals, such as PZNT and PMNT offer significant electromechanical coupling (k_{33}>90%) in contrast to Pb(Zr, Ti)O3(PZT) polycrystalline ceramics (k_{33}~75%), making them promising candidates for medical ultrasound transducers. However, their relatively low Curie temperature (Tc ~150-170◦C) and their substantially lower ferroelectric (rhombohedral to tetragonal) phase transition temperatures(T_{R-T})~60–120◦C), may limit their implementation in transducers in which thermal stability is desired in terms of dielectric and piezoelectric property variation and depolarization as a result of postfabrication processes. [13-16]

To achieve high performance over a wide temperature range, many recent advances focused on the study of relaxor-PT systems and MPB compositions in the $Bi(Me)O_3$-$PbTiO_3$ (Me=Sc^{3+}, In^{3+}, Yb^{3+}) family in the single crystals had been achieved. Based on a perovskite tolerance factor Tc relationship, the new family of MPB $Bi(Me)O_3$-$PbTiO_3$ systems has been shown to exhibit Tc's>450°C in the polycrystalline ceramics. New crystals with higher Curie temperatures are desired for the future piezo/ferro-electrical device applications.

4.1. THE DEVELOPMENT OF TYPICAL HIGH CURIE TEMPERATURE $Pb(B_1B_2)O_3$-$PbTiO_3$ FERROELECTRIC CRYSTAL

Yasuda et al. reported single crystals in the system $Pb(In_{1/2}Nb_{1/2})O_3$– $PbTiO_3$ (PINT) with a Curie temperature of 260°C and a piezoelectric coefficient d_{33} of 700pC/N. Furthermore, perovskite single crystals in the (1-x)$Pb(Yb_{1/2}\ Nb_{1/2})O_3$–x$PbTiO_3$ (PYNT) system have been studied. It was found that near the MPB, the Curie temperature was around 350°C and the piezoelectric coefficient was up to 2500pC/N. [17-18]

To elevate the Curie temperature of $Pb(B_1, B_2)O_3$ systems, the route of dopant modifications and ternary systems had been adopted, such as the development of the crystals of Fe-doped 0.955$Pb(Zn_{1/3}Nb_{2/3})O_3$-0.045$PbTiO_3$ (Fe-PZNT4.5, 0.41Pb(In1/2Nb1/2)O_3–0.17$Pb(Mg_{1/3}Nb_{2/3})O_3$–0.42$PbTiO_3$ [PIMNT(41/17/42)], $Pb(In_{1/2}Nb_{1/2})O_3$-$Pb(Mg_{1/3}Nb_{2/3})O_3$-

$PbTiO_3$ (PIN-PMN-PT) and $0.34Pb(In_{1/2}Nb_{1/2})$ O_3–$0.25Pb(Mg_{1/3}Nb_{2/3})O_3$–$0.41PbTiO_3$ (34PIN-25PMN-41PT) systems. [19-25]

In Fe-PZNT4.5 single crystals, the Curie temperature and the rhombohedral to tetragonal phase transition temperature ($T_{r\text{-}t}$) were about 143 and 92 °C, respectively, more than 20 °C lower than that of the pure crystals. The room temperature dielectric permittivity along the <001> direction after poling was about ~6000 (5000-7000), increased compared to that of pure PZNT4.5 crystals ~3000-5000. The remnant polarization (P_r) and coercive field (E_c) of <001>-oriented doped crystals measured at 1 Hz and 20 kV/cm field were ~27.3 $\mu C/cm^2$ and 4.7 kV/cm, respectively. The latter was significantly greater than the E_c of the pure crystals ~3 kV/cm. Piezoelectric coefficients (d_{33}) of <001>-oriented doped PZNT4.5 crystals were ~2500-3000 pC/N, which was much higher than that of the pure (1400–2000 pC/N). [22]

Tetragonal PIMNT(41/17/42) ferroelectric phase with relatively high Curie temperature of 253°C, exhibiting the relative permittivity of 487 and low dielectric loss of 0.3% at 50 Hz and room temperature satisfied the needs of operation as a high Curie temperature material. The PIMNT crystal with better temperature stability is a good candidate as an infrared detector material compared with the pure PMNT single crystals. [23-24]

Many studies have been carried out to enhance the Curie temperature in the relaxor-based single crystal family, such as $Pb(In_{1/2}Nb_{1/2})O_3$– $PbTiO_3$ and $Pb(Yb_{1/2}Nb_{1/2})O_3$–$PbTiO_3$ single crystals, in which Curie temperatures are about 260 and 350 °C near the morphotropic phase boundary, respectively.

Luo et al. grew the high Cure temperature and larger coercive field ternary $Pb(In_{1/2}Nb_{1/2})O_3$-$Pb(Mg_{1/3}Nb_{2/3})O_3$-$PbTiO_3$ (PIN-PMN-PT or PIMNT) single crystals, which exhibited the comparable piezoelectric properties with PMN-PT crystals. The 1.6MHz ultrasonic transducers based on these single crystals can be used in transcranial Doppler (TCD) ultrasonography. The -6dB bandwidth of PIN-PMN-PT single crystal based transducer is nearly 10% broader.

The ternary PINPMN-PT crystals, promising candidates for their higher MPB temperature ($Trt \geq 100$ °C), Curie temperature ($TC \geq 150$ °C) and coercive field ($EC \geq 5$ kV/cm), were grown directly from their melt by the modified Bridgman technique. The single crystals PIN-PMN-PT were grown and investigated for medical ultrasonic transducer applications and the dielectric, piezoelectric and uniformity of PIMNT were investigated. Single element transducers based on crystal were fabricated.

The medical ultrasonic transducers require improvement of the sensitivity and bandwidth of crystal-based probe. Due to the enhanced piezoelectric property of PIN-PMN-PT single crystals, the ultrasonic phased array based on these materials had achieved the significant improvements. The 4dB larger sensitivity, 40% shorter impulse response and 20% broader bandwidth (-6dB) of PIN-PMN-PT single crystal-based will improve medical imaging quality. The large size crystals with ultrahigh piezoelectric properties and good thermal stability may become the promising candidate for medical ultrasonic transducers applications.

S. Zhang et al had reported the shear-mode piezoelectric and electromechanical properties of $0.55Pb(Yb_{1/2}Nb_{1/2})O_3$-$0.45PbTiO_3$ (PYNT45) single crystals. The piezoelectric coefficients d_{15} and g_{15} can reach to 2100 pC/N and 0.064 Vm/N, respectively, and electromechanical coupling coefficient $k15$ can also reach to 91% at room temperature. Above 168 °C, the values decreased dramatically when increasing temperature. The ferroelectric– ferroelectric phase transition temperature can be estimated to be 168 °C. High temperature aging experiments indicated these single crystals exhibit good domain stability.

4.2. THE DEVELOPMENT OF $BiScO_3$-$PbTiO_3$ SINGLE CRYSTAL [25-30]

MPB compositions have anomalously high dielectric and piezoelectric properties as a result of enhanced polarizability arising from the coupling between two equivalent energy states, i.e., the tetragonal and rhombohedral phases, allowing optimum domain reorientation during the poling process. Further modifications using acceptor and donor dopants give us the wide range of piezoelectric compositions.

Generally, three approaches can be exploited to enhance the temperature usage of perovskite ferroelectric crystals. The first way is to make the dopant modifications (~1 mole%) to PZNT and PMNT systems to elevate Tc and $T_{R\text{-}T}$. The second way is to develop MPB in $Pb(Yb_{1/2}Nb_{1/2})O_3$– $PbTiO_3$ (PYNT) crystals, which possesses the highest Tc for a relaxor-PT system (on the order of ~350◦C). Third, is the development of novel MPB compositions Bi(Me)O3–PbTiO3 (Me = Sc3+, In3+, Yb3+) family crystals. Based on a perovskite tolerance factor Tc relationship, this new family of MPB systems has been shown to exhibit Tc> 450◦C in polycrystalline ceramics. Specially, a significantly high level of piezoelectric activity and excellent temperature stability with a Tc ~ 460◦C, are anticipated for the single crystals in BSPT (Me = Sc3+ (BSPT)).

For the B-site modifications , including Fe3+, Ni2+, W6+, Mo6+, Zr4+, etc., the $T_{R\text{-}T}$ and Tc of PZNT and PMNT single crystals were reduced on the order of 10–30◦C, although the dielectric and piezoelectric properties were intrinsically enhanced. For instance, the Fe-doped PZNT4.5 single crystals possessed higher piezoelectric and electromechanical properties than those of pure PZNT4.5 crystals, but the $T_{R\text{-}T}$ shifted down to only 85◦C. For the A-site modified PMNT crystals. the temperature curve of the modified PMNT crystal was effectively shifted up about 10◦C for both Tc and $T_{R\text{-}T}$, while the piezoelectric properties were still comparable to that of the pure one.

A Tc ~330◦C was observed in the high Tc of PYNT crystals near the MPB composition, being significantly higher than any other relaxor-PT system. A dielectric anomaly also was observed at ~160◦C, corresponding to a $T_{R\text{-}T}$ phase transition, reflecting once again the strong curvature near the MPB composition. With a $T_{R\text{-}T}$ is nearly half of its Tc, these PYNT crystals offer a greater range of temperature usage than that for PZNT and PMNT, while maintaining high levels of piezoelectric activity.

A series of doped and novel BSPT perovskite compositions were grown from the high temperature flux method. The Pb3O4 or PbO-Bi2O3 was chosen as the flux and opant modifications were fixed at ~1 mole%. The typical size of the grown crystal size for PYNT and BSPT single crystals is on the order of 5–10 mm. The as-grown crystals were oriented along the pseudocubic crystallographic axis [001] and [110]. [25-30]

Rhombohedral Single Crystals of 0.43BiScO3– 0.57PbTiO3

The 0.43BiScO3– 0.57PbTiO3 (BSPT57) perovskite single crystal solid solution possesses a high Curie temperature (*TC:* 402 °C) and rhombohedral to tetragonal phase transition temperature ($T_{R\text{-}T}$:349 °C), with excellent electromechanical properties (*k*33: 90%). [29]

The BSPT57 single crystals were grown by the high temperature solution method using PbO and Bi2O3 as self-flux. The samples used for resonance measurements were oriented along their crystallographic directions (001) and sputtered with gold electrodes. A high temperature of 120 °C and high electric field of 30 kV/cm were applied on the samples for 10 minutes for poling the samples.

The resonance method was used to determine all of the eleven electromechanical coefficients of the BSPT57 piezocrystals using five *Z*-cut samples. The complete set of material constants, including the elastic s_{ij} , c_{ij} , piezoelectric d_{ij} , e_{ij} , g_{ij} , h_{ij} , dielectric ε_{ij} , and electromechanical k_{ij} values were derived from the experimental data. The BSPT57 crystal was found to be a promising material for piezoelectric applications, which has a much high temperature stability, but is comparable to the properties of the $Pb(Zn_{1/3}Nb_{2/3})O_3$– $PbTiO_3$ and $Pb(Mg_{1/3}Nb_{2/3})O_3$– $PbTiO_3$ relaxor-PT single crystals.

From the measured and derived complete set of material coefficients for BSPT57 single crystals, one can see that the BSPT57 single crystal possesses high electromechanical coupling factors and piezoelectric coefficients, compared to the materials coefficients of PZNT and PMNT single crystals. BSPT crystals also have lower piezoelectric strain coefficients for longitudinal and lateral modes but higher values for the shear mode piezoelectric voltage coefficient. The shear electromechanical coupling factor k_{15} was found to be 59%, much higher than the values for PZNT and PMNT crystals (k_{15}:23%–32%), indicating excellent shear mode properties of the BSPT crystals, which is promising for possible sensor applications.

Because of the anisotropy of the crystals, the lateral coupling factor k_{31} is different when the vibration orientation is along (100) or (110) for BSPT. For BSPT57, a much higher coupling factor when vibrated along the (110) direction was found. The piezoelectric coefficients along different directions are almost the same because the elastic compliance along the (110) direction is much smaller than that of (100) direction.

BSPT57 single crystals were characterized by using the resonance technique. A complete set of elastic, piezoelectric, electromechanical, and dielectric constants were obtained and which showed that BSPT crystals possessed excellent electromechanical properties.

The room temperature dielectric permittivity of (001)-oriented BSPT (43/57) crystals was found to be ~3000, with a dielectric loss of 4% after the poling process. The cure of dielectric permittivity and loss as a function of temperature were measured from room temperature to 500 °C, by using a multifrequency LCR meter (HP4284A). A dielectric maximum was observed at 404 °C (the Curie temperature). A second dielectric anomaly was observed at 349 °C, according to the R–T ferroelectric phase transition. In the range from room temperature to 300 °C, the dielectric permittivity is nearly temperature independent, implying the thermal stability associated with a high T_C piezoelectrics. In the temperature range of 150 to 350 °C, the dielectric loss decreased to 2%, but reached a peak at the Curie temperature.

The hysteresis and strain-field loops (001)-oriented BSPT(43/57) crystals were measured by using a modified Sawyer–Tower circuit driven by a lock-in amplifier (SR830) at 1 Hz and 20 kV/cm electric fields, from which the coercive field (*Ec*) was found to be 13.7 kV/cm, and the remnant polarization was 23 mC/cm^2. The unipolar strain behavior exhibited low hysteresis, with strains reaching 0.12% at 10 kV/cm. With this curve, the piezoelectric coefficient d_{33} can be calculated directly to be 1200 pC/N, similar to that measured by using a Berlincourt *d*33 meter. A maximum strain of 1.1% was reached under 30 kV/cm, as the crystal transformed from the R to T state with large hysteresis. However, the coercive field is significantly higher than that for PZNT and PMNT systems, being on the order of 2–3 kV/cm.

The electromechanical coupling factor of a (001)-oriented BSPT(43/57) can be determined by the resonance (fr) and antiresonance (fa) frequencies method, by using an impedance analyzer (HP4194A), from which the longitudinal electromechanical coupling factor $k33$ was calculated to be 90%; and the calculated piezoelectric coefficient d_{33} of 1150 pC/N was consistent with the values obtained from the unipolar strain measurement.

In the range of 25 to 320 °C, k_{33} decreased a little (to 89%) while the temperature increased. Above 330 °C, the coupling factor decreased to 88.5%, and further decreased to only 83% when the temperature reached 340 °C, corresponding to the value of coupling factor for BSPT tetragonal crystal. The same tendencies for the piezoelectric coefficients were also observed. This sudden drop above 340 °C in the coupling factors and piezoelectric coefficients can be ascribed to the R–T phase transformation. The various dielectric, piezoelectric, and electromechanical property characterizations of BSPT(43/57) crystals and relaxor-PT crystals and polycrystalline BSPT ceramics are presented.

The full set of material constants for BSPT57 crystals are helpful for both the theoretical and practical exploitation of this material. Together with its very high Curie temperature and rhombohedral–tetragonal phase transformation temperature and large coercive field, BSPT57 crystals are promising materials for high temperature high performance applications.

BSPT58 Crystals

From the dielectric temperature behavior of the crystals of BSPT58 with PT content at 58% near the MPB composition, [30] a Tc of 410◦C was found, being comparable to that reported in polycrystalline ceramics. A weak dielectric anomaly was also observed at ~357◦C, being associated with a rhombohedral-tetragonal phase transformation. Compared to BSPT57 crystals, both the Tc and T_{R-T} for BSPT58 crystals exhibit 8◦C higher, and the dielectric and piezoelectric values were slightly increased, showing better properties than BSPT57 crystals.

The longitudinal piezoelectric coefficients and coupling factors for BSPT58 crystals as a function of temperature indicated that, the temperature stability of this system is far superior to the other perovskite crystals, maintaining a coupling coefficient k_{33} ~91% up to 200◦C, which slightly decreases to 89% up to 350◦C, corresponding to the phase transition temperature T_{R-T}, which is decreased to 82% when the temperature is over T_{R-T}. The piezoelectric coefficients increased from 1300 pC/N at room temperature to 3100 pC/N at 350◦C. The same tendency was observed for the lateral properties for BSPT58 crystals, with the lateral coupling factor $k31$ found to be 80% at room temperature being nearly constant till 350◦C, while the lateral piezoelectric coefficient $d31$ increased from −670 pC/N at room temperature to −1500 pC/N at 350◦C.

Tetragonal BiScO3-PbTiO$_3$ Single Crystal [28-30]

Compared to rhombohedral phase single crystals, tetragonal crystals showed the advantage of a higher Curie temperature and no phase transition temperature T_{R-T} was observed below the T_C. They also exhibited higher temperature stability of the properties, while still keeping a relatively high electromechanical coupling factor.

More recently, a new perovskite system $BiScO_3$-$PbTiO_3$ (BSPT) was obtained in both ceramic and single crystal forms. The BSPT materials with a MPB composition (0.36BS-0.64PT) showed a high Curie temperature at 450 °C, which is 100 °C higher than the commercial PZT materials, and a better piezoelectric property compared to PZT ceramic.

Modified $BiScO_3$-$PbTiO_3$ (BSPT) tetragonal single crystals were grown using the high temperature solution method. Bi_2O_3, Sc_2O_3, Pb_3O_4 and TiO_2 powders with high purity were selected as starting materials and Pb_3O_4 with a special ratio to Bi_2O_3 was used as flux. The X-ray diffraction was performed on the grounded crystal powder. The samples used in dielectric and piezoelectric measurements were oriented along its crystallographic [001] direction using real-time Laue system and gold electrodes were sputtered to the parallel (001) faces.

The dielectric, piezoelectric and elastic properties of single domain BSPT crystals, after poling along [001] crystallographic direction, have been determined experimentally using the resonance method. The results showed that the BSPT tetragonal crystals possess good piezoelectric properties, with an electromechanical coupling factor of about 88% and piezoelectric coefficient over 400 pC/N at room temperature. BSPT tetragonal crystals have a high Curie temperature around 436 °C and a high coercive field of 28 kV/cm. The crystal exhibited very good temperature stability of the properties up to 380 °C. This makes BSPT tetragonal crystals a promising candidate for some high temperature actuator and sensor applications.

0.34$BiScO_3$–0.66$PbTiO_3$ Single Crystals [30]

Novel tetragonal phase perovskite ferroelectric (1x)$BiScO_3$–x$PbTiO_3$ single crystals with a typical size of 2–15mm and with rectangular shape and green/blue color, were successfully grown by a conventional high temperature solution method using a Pb_3O_4 and Bi_2O_3 flux. The dielectric constant (K) at room temperature was determined to be 400–550, decreasing to about 300 after the poling process. The Curie temperature of 0.43$BiScO_3$–0.66$PbTiO_3$ single crystals was found to be about 460 °C, similar to the polycrystalline data. The remanent polarization (Pr) was 28 mC/cm^2 with a coercive field (Ec) of 34 kV/cm. The piezoelectric coefficient d_{33} was about 200 pC/N for 001 oriented tetragonal crystals. The longitudinal electromechanical coupling coefficient k_{33} was 73% with a lateral mode coupling coefficient k31 of 35.8% and a thickness mode coupling kt of 60–64%.

The BSPT crystal with a composition (x=0:66) near the MPB, was grown using a flux method. The starting materials were high-purity powders Pb_3O_4, Bi_2O_3, Sc_2O_3 and TiO_2, among which Pb_3O_4 and Bi_2O_3 with ratios according to the PbO–Bi2O3 phase diagram were selected as the flux, owing to their low melting temperature and chemical compatibility with the stability compositions. The BSPT polycrystalline material and flux were loaded into platinum (Pt) crucibles, which was sealed by a Pt foil to minimize the evaporation of the flux. After the Pt crucibles were put into an alumina crucible sealed with alumina lid and alumina cement, the growth was carried out in a muffled resistance furnace. The materials were heated to 1200 °C and soaked for 7 hours, followed by slow cooling (0.1–5 °C/h) to 900 °C. The crucible was furnace-cooled to room temperature.

The as-grown single crystals of BSPT were green/blue in color and transparent with 2–15mm in size. The main defects in the BSPT single crystals are flux inclusions and (0 0 1) cleavage planes. The X-ray powder diffraction (XRD) measurement indicated that the crystals are tetragonal. The X-ray diffraction on the emerged faces revealed that the habitual faces of the obtained crystal are {0 0 1}. The c/a ratio of BSPT single crystal is 1.03, which is larger than that of PZNT, corresponding to a BSPT66 for polycrystalline materials just on the tetragonal side of MPB. The density was measured to be 7.5 g/cm^3, very close to the polycrystalline materials.

The room temperature dielectric constant varied between 400 and 550, with the dielectric loss about 0.5%. After high-temperature poling, the dielectric constant decreased to about 300. From the dielectric constant and loss temperature dependence curves at 10 and 100 kHz frequencies, the dielectric maximum occurred at about 460 °C (Tc, Curie temperature), which is much higher than the Tc of PZNT and PMNT perovskite crystals (about 150–180 °C). Compared with the Tc of the BSPT polycrystalline materials, the PT content in the tetragonal BSPT single crystals is about 66%. The dielectric loss tangent also showed a maximum value in the vicinity of Curie temperature and increased rapidly with the temperature increasing, which is associated with a conductive loss.

The measured polarization-electric field hysteresis loop of a BSPT single crystal shows a remanent polarization (Pr) of 28 mC/cm2 and coercive field (Ec) of 34 kV/cm. Strain-field data, for the same measured conditions, follow the classical bipolar ferroelectric switching behavior, displaying a symmetric butterfly loop. A constrained pinched P–E hysteresis behavior and strain behavior were similar to that of 'hard' piezoelectric ceramics. This the constrained loop ascribed to a defect dipole created by an association between an acceptor charge and a compensate oxygen vacancy, pinning the walls, and preventing nucleation and growth of domains in the direction favored by an applied electric field.

The strain-field curve measured at 60 kV/cm for a (001)-oriented sample shows the strain-E field hysteresis is very low with a d_{33} of 200 pC/N, similar to the value obtained using a Berlincourt d_{33} meter.

Tetragonal BSPT may be a good candidate for the single-element transducer in ultrasound application. The rhombohedral BSPT crystal should exhibit high E-induced strain and high Curie temperature owing to a large c/a ratio of 1.03, which makes it a promising material for the next generation high temperature and high performance actuator and transducer applications.

$xBiScO_3$–$yBiGaO_3$–(1xy)$PbTiO_3$ Single Crystals [27]

The $xBiScO_3$–$yBiGaO_3$–(1xy)$PbTiO_3$ (BS–BG–PT) system single crystals were grown by the high temperature solution method using Pb_3O_4 and Bi_2O_3 as the flux. The dielectric permittivity ε at room temperature for unpoled tetragonal crystals was determined to be 500–600 with dielectric loss tangents less than 0.3%. The Curie temperature was in a band of 420–450 °C, with a dielectric maximum, exhibiting relaxor behavior. The d_{33} was 300 pC/N for (001 oriented tetragonal crystals with electromechanical coupling factor (k_{33}) of 75%, with a shear mode, d_{15} of 290 pC/N and k_{15} of 45%; lateral mode, d_{31} of 55 pC/N and k_{31} of 37%. The remnant polarization (Pr) was 46 mC/cm^2 with a coercive field (Ec) of 43 kV/cm.

The linear E-O coefficient r_{33} and r_{13} were 36 and 4 pm/V at room temperature, close to the values of the LN crystal. But BS–BG–PT crystals have much higher piezoelectric coefficients and electromechanical factors. All the dielectric permittivity, electromechanical coupling factors and piezoelectric coefficients, together with the E-O coefficients, showed much better temperature stability to 200 °C compared to PZNT and PMNT single crystals, indicating that the crystal is a promising material in both electric, E-O and acoustic-optic devices for high temperature applications.

The dielectric, piezoelectric, electromechanical and linear E-O properties of BS–BG–PT tetragonal single crystals are listed in Table 4.1.

Regardless of the composition in order of increasing transition temperature (Tc and $T_{R\text{-}T}$), domain engineered [001] rhombohedral relaxor-PT crystals offer high levels of piezoelectric activity with d_{33}'s > 1200 pC/N and electromechanical coupling factor *k*33 on the order of 90% for all systems. Attempts to expand the phase transition temperatures using minor dopant modification in PZNT and PMNT crystals were limited with B-site additions, generally reducing both Tc and TR-T. Although the highest Tc relaxor-PT crystal system known, PYNT crystals exhibited a TR-T of only ~160◦C, which is still much lower than commercial PZT ceramics.

Table 4. . The detailed properties of the BS-BG-PT tetragonal single crystals

Orientation	T_c (°C)	$\varepsilon_{33}^T/\varepsilon_0$	$\varepsilon_{11}^T/\varepsilon_0$	$\tan\delta$	P_r (μC/cm²)	P_s (μC/cm²)	E_c (kV/cm)	d_{33} (pC/N)
⟨001⟩	420–450	500–600	1550	0.3%	46	50	43	300
d_{31} (pC/N)	d_{15} (pC/N)	k_{33}	k_{31}	k_{15}	r_{33}^* (pm/V)	r_{13}^* (pm/V)	r_{33} (pm/V)	r_{13} (pm/V)
−55	290	75%	−37%	45%	47	19	36	4

Table 4.2. The comparison of the properties of several high Curie relaxor based ferroelectric single crystals

Crystal	Tc (°C)	$T_{R\text{-}T}$ (°C)	Pr (μC/cm²)	E_C (kV/cm)	K_{33}^T	d_{33} (pC/N)	k_{33}	d_{31} (pC/N)	k_{31}
PZNT4.5	140	85		3.3	5200	2500-3500	93%	-1000	84%
Fe-PZNT4.5	143	92	27.3	4.7		2500-3500			
PMNT30	140	97		2	5100	2000	90%	-800	80%
PYNT40	270	168		10	2700	1200	88%	-500	76%
PYNT45	325	160		12.5	2000	1200-2000	88-90%	-550	78%
PIMNT (41/17/42)	253								
PIN-PMN-PT	160-200	120-130		5.5		1100-1500	89-92%		
BSPT66	460		28	34		200	73%	-44	-35.8%
BSPT57	402	349	23	13.7	3000	1150	90%		
BS-BG-PT	420~450		46	43		300	75%	-55	-37%
Modified BSPT	436			28			87.8%	-162	51.8%

Table 4.3. The growing method and the size of several high curie relaxor based ferroelectric single crystals

Materials	Researcher	Method	Size(mm)
PYNT	N.Yasuda	Flux	1~2
	Zhang Shujun	Flux	2~8
BSPT	Zhang Shujun	Flux	2~15
BS-BG-PT	Zhang Shujun	Flux	3~8

(Flux: high flux method. Bridgeman method.)

The new bismuth-based perovskite MPB system BSPT was found to possess a competitive temperature usage range to PZT ceramic (>100◦C higher) while exhibiting the comparable piezoelectric properties as PZNT and PMNT single crystals, which offer great potential for the next generation of high temperature, high performance transducer materials.

For indicating clearly the development of the High-Tc ferroelectric crystals, Table 2 shows the comparison of the properties of several high Curie relaxor based ferroelectric single crystals. It reveals PYNT, BSPT and BS-BG-PT with the high Cutie temperature. The dopant modifications and ternary systems may enhance the Curie temperature to a certain extent. To achieve high performance with high levels of piezoelectric activity over a wide temperature range, many methods can be used to develop more effective materials and satisfy the needs of modern industry.

However, there are many problems with the practical application of these high curie temperature ferroelectrics, such as the difficult growth crystals and the small size of crystals, shown in Table 3. Because of the limit to the crystal size, the performance of crystals cannot be completely achieved.

REFERENCES

[1] B. Jaffe, *J. Res. Natl. Bur. Stand.* 55 (1955) 239.

[2] S. Park and T. Shrout, "Characteristics of relaxor-based piezo- electric single crystals for ultrasonic transducers," *IEEE Trans. Ultrason., Ferroelect., Freq. Contr.*, vol. 44, pp. 1140–1147, Sep. 1997.

[3] S.-E. Park and T. R. Shrout, "Ultrahigh strain and piezoelectric behavior in relaxor based ferroelectric single crystals," *J. Appl. Phys.*, vol. 82, pp. 1804–1811, Aug. 1997.

[4] S. Saitoh, T. Takeuchi, T. Kobayashi, K. Harada, S. Shimanuki, and Y. Yamashita, "An improved phased array ultrasonic probeusing Pb(Zn1/3Nb2/3)O3–PbTiO3 single crystal," *Jpn. J. Appl. Phys.*, vol. 38, pp. 3380–3384, May 1999.

[5] S. J. Zhang, L. Lebrun, S. F. Liu, S. Rhee, C. A. Ran- dall, and T. R. Shrout, "Piezoelectric shear coefficients of Pb(Zn1/3Nb2/3)O3–PbTiO3 single crystals," *Jpn. J. Appl. Phys.*, vol. 41, pp. L1099–1102, Oct. 2002.

[6] H.S. Luo, G.S. Xu, H. Q. Xu, P.C.Wang, andZ.W. Yin, "Compositional homogeneity and electrical properties of lead magnesium niobate titanate single crystals grown by a modified Bridgman technique," *Jpn. J. Appl. Phys.*, vol. 39, pp. 5581– 5585, Sep. 2000.

[7] S. J. Zhang, L. Lebrun, D. Y. Jeong, C. A. Randall, Q.M. Zhang, and T. R. Shrout, "Growth and characterization of Fe-doped Pb(Zn1/3Nb2/3)O3–PbTiO3 single crystals," *J. Appl. Phys.,* vol. 93, pp. 9257–9262, Jun. 2003.

[8] R. Eiter, C. A. Randall, T.R. Shrout, P. Rehrig, W. Hackenberger, and S.-E. Park, "New temperature morphotropic phase boundary piezoelectrics based on the Bi(Me)O3-PbTiO3 ceramics," *Jpn. J. Appl. Phys*., vol. 40, pp.5999-6002, Oct. 2001.

[9] S.-E. Park, T.R. Shrout, *J. Mater. Res. Innov*. 1 (1997) 20.

[10] S.-E. Park, T.R. Shrout, *Jpn. J. Appl. Phys*. 36 (1997) 1154.

[11] S.-E. Park, T.R. Shrout, *J. Appl. Phys*. 82 (1997) 1804.

[12] S. Shimanuki, S. Saito, Y. Yamashita, *Jpn. J. Appl. Phys*. 37 (1998) 3382.

[13] M. Dong, Z.-G. Ye, *J. Crystal Growth* 209 (2000) 81.

[14] H.S. Luo, G.S. Xu, Z.W. Yin, *Jpn. J. Appl. Phys*. 39 (2000) 5581.

[15] S.-E. Park, T.R. Shrout, *J. Appl. Phys*. 82 (1997) 1804.

[16] H.S. Luo, G.S. Xu, Z.W. Yin, *Jpn. J. Appl. Phys*. 39 (2000) 5581.

[17] N. Yasuda, H. Ohwa, M. Kume, Y. Yamashita, *J. Crystal Growth* 229 (2001) 299.

[18] N. Yasuda, H. Ohwa, Y. Yamashita, *Jpn. J. Appl. Phys*. 39 (2000) 5586.

[19] S.J. Zhang, P.W. Rehrig, C.A. Randall, T.R. Shrout, *J. Crystal Growth* 234 (2002) 415.

[20] S.J. Zhang, C.A. Randall, T.R. Shrout, *Jpn. J. Appl. Phys*. 41 (2002) 2A.

[21] S. J. Zhang, L. Lebrun, D. Y. Jeong, C. A. Randall, Q.M. Zhang, and T. R. Shrout, "Growth and characterization of Fe-doped Pb(Zn1/3Nb2/3)O3–PbTiO3 single crystals," *J. Appl. Phys*., vol. 93, pp. 9257–9262, Jun. 2003.

[22] Growth and pyroelectric properties of high Curie temperature relaxor-based ferroelectric Pb(In1⁄2Nb1⁄2)O3-Pb(Mg1⁄3Nb2⁄3)O3–PbTiO3 ternary single crystal, *Applied Physics Letters* 92, 252907 (2008), Peng Yu.

[23] Characterization of Pb„In1/2Nb1/2…O3–Pb„Mg1/3Nb2/3…O3–PbTiO3 ferroelectric crystal with enhanced phase transition temperatures, *Journal of Applied Physics* 104, 064106 (2008), Shujun Zhang.

[24] Temperature and electric field dependence of the dielectric property and domain evolution in [001]-oriented 0.34Pb(In1/2Nb1/2)O3–0.25Pb(Mg1/3Nb2/3)O3–0.41PbTiO3 single crystal, *Journal of Applied Physics* 109, 014111(2011).

[25] Eitel Richard E, Randall Clive A, Shrout Thomas R, et al. *Jpn J Appl Phys*, 2001, 40(10):5999.

[26] S. J. Zhang, L. Lebrun, S. Rhee, R. E. Eitel, C. A. Randall, and T. R. Shrout, "Crystal growth and characterization of new high Curie temperature (1-x)BiScO3–PbTiO3 single crystals," *J. Cryst. Growth*, vol. 236, pp. 210–216, Mar. 2002.

[27] S. J. Zhang, D. Y. Jeong, Q. M. Zhang, and T. R. Shrout, "Electromechanical and electro-optic properties of xBiScO3–yBiGaO3–(1-x-y)PbTiO3 single crystals," *J. Cryst. Growth*, vol.247, pp. 131–136, Jan. 2003.

[28] S. J. Zhang, Clive A. Randall, and Thomas R. Shrout, "High Curie temperature piezocrystals in the BiScO3–PbTiO3 perovskite system," *Appl. Phys. Lett*., vol. 83, pp. 3150–3152, Oct. 2003.

[29] Characterization of perovskite piezoelectric single crystals of $0.43BiScO_3$–$0.57PbTiO_3$ with high Curie temperature, *J. Appl. Phys.*, Vol. 95, No. 8, 15 April 2004, S. J. Zhang, Clive A. Randall, and Thomas R. Shrout.
[30] Dielectric, piezoelectric and elastic properties of tetragonal $BiScO_3$-$PbTiO_3$ single crystal with single domain, S. Zhang et al. / *Solid State Communications* 131 (2004) 41–45.

Chapter 5

HIGH-TC FERROELECTRIC THIN FILMS

High ferroelectric and piezoelectric performance and good temperature stability is very important in device applications, especially in automotive, aerospace, and related industries. $BiScO_3$-$PbTiO_3$(BSPT) thin films with a high T_c are attractive. The high Curie temperature piezoelectric BSPT thin films and the recent progress with these materials is discussed. The mechanical properties of sol-gel derived BSPT thin films annealed at different temperatures were investigated by nanoindentation. The main properties of the homogeneous, up-graded, and down-graded thin films were compared. The effects of different factors on the sol–gel-derived BSPT thin films were presented. The piezoresponse data as a function of thickness correlates with the structural modification of thin film.

5.1. FERROELECTRIC THIN FILMS BY PULSED LASER DEPOSITION AND SOL-GEL TECHNIQUES: PROCESSING AND CHARACTERIZATION

In the last few decades, ferroelectric thin films and their applications have gained considerable attention, since they possess excellent properties such as large remnant polarization, large dielectric constant, good pyroelectric and piezoelectric effect and good electro-optic effect. There are many methods for preparing thin films based on both physical and chemical principles. The physical methods can be divided into thermal evaporation and sputtering, and the chemical methods into vapor phase and liquid phase. [1-4]

Ferroelectric thin films, such as $BaSrTiO_3$, $PbTiO_3$, Pb(ZrxTi1-x)O_3, $SrBiTaO_3$, and $Bi_4Ti_3O_{12}$ films, can be used in many electronic device applications, in which the applications of ferroelectric thin films use the unique dielectric, piezoelectric, pyroelectric, and electro-optic properties of ferroelectric materials. These applications include: non-volatile memories using the high polarizability; thin film capacitors using the dielectric properties; and pyroelectric sensors using the changing of the dielectric constant with temperature; and piezoelectric actuators using the piezoelectric effect, etc.

Due to the possibilities provided to replace the current CMOS memory used for Random Access memories with non-volatile Ferroelectric Random Access Memories (FRAM), the perovskite structured ferroelectric thin films have been given a lot of attention in the last few years. Although then low density FRAMs have been available on the commercial market, the high-density FRAMs are now in demand and being developed. For example, Korean

HYUNDAI Electronic Industries are now developing 4Mb FRAMs based on Strontium Bismuth Tantalum (SBT).

For the preparation and characterization of the ferroelectric thin film capacitors, two Au/PZT/LSCO/Al_2O_3 thin film capacitors have been prepared using two different techniques: Pulsed Laser Deposition (PLD) and spin coating sol-gel technique. The third Au/PZT/Pt/Si/O_2/Si thin film capacitor was made by the spin coating sol-gel technique. *Pole figure plots* have been performed to study the films' orientation. Cross-sectional SEM micrographs have been made to see what differences there are between the two deposition techniques at the microstructure inside the films, and AFM has been used to study the microstructure of the surface. A description of the two deposition techniques and the *pole figure plots* are given in the experimental section.

5.1.1. Memory Technology [5-8]

The memories used today can be grouped into two categories: *volatile* and *non-volatile*. These two memory technologies are currently dominating the market. The volatile types are memories that contain information which has to be changed often, and thus easy to write. They are RAMs using the capacitance of a dielectric and the semiconductivity of silicon, such as Dynamic Random Access Memories (DRAM) and Static Random Access Memories (SRAM). These memories have the disadvantage that they *are* volatile and lose the stored information when the power is turned off. This is a problem, for example, during power failures or an insufficient battery power situation.

Today, there exist non-volatile memories such as Electrically Erasable Read Only Memories (EEPROM) and Complementary Metal Oxide Semiconductors (CMOS) with battery backup. The non-volatile memories are used as *read only* or *read mostly memories*, due to the difficulty to write. They also had the disadvantage that they have very long write cycles - from some milliseconds to several microseconds. In harsh environments where the long writing cycle makes the possibilities of noise disturbance much higher, this is unfavorable. They need a high writing voltage (>12 V) and have a low writing endurance. Compared with EEPROM, the FRAM technology has many benefits. It has a much shorter write cycle and a higher endurance which makes the FRAMs very good for data collection and log applications.

Although the non-volatile FRAMs are not able to overtake the DRAMs leading market position, for the higher endurance of the DRAM, the shorter write cycle (~60 ns) and the higher density plus the fact that the industry already has a process line for these devices. The FRAM technology is growing rapidly because of the benefits of having non-volatile memories. There are two ways to use the ferroelectricity in memory devices. The first way is to use the ferroelectricity of the capacitor to pole the memory cell, and thus give it a value as either 0 or 1. The readout of the memory cell is done by applying an E and measuring the switching charges from the capacitor together with a reference capacitor and thereby decide the value of the memory cell. This is the basic DRAM design and it has a destructive readout. The other way is the Ferroelectric Field Effect Transistor (FeFET). In this way, the poling of the material is used to deplete an semiconductor, which gives a nondestructive direct readout, and makes this device faster than the one previously mentioned. This implies that the memory technology based on ferroelectricity will be of greater importance in the future.

5.1.2. Thin Film Characteristics [9-11]

The following properties of the thin film are important for memory applications: remnant polarization, coercive field, fatigue, retention, imprint, tanδ, switching speed, grain size, film thickness and the compatibility between the thin film material and the rest of the integrated circuit.

It is necessary for the hysteresis loop to show a large remnant polarization and a low coercive field. This can be obtained by composition and substitution control during the film process. For instance, introducing BiTi(TaNb) to SBT at ratio 0.7 SBT /0.3 BTT will give rise to a better Pr. The small film thickness is required to keep the switching voltage low. In the future, a thickness less than 60 nm is probably needed to decrease the voltages to 1 volt. At such a low voltage, the film quality is of great importance, and the fatigue of the ferroelectric thin film must be controlled.

Many studies have been made on the fatigue of ferroelectrics, and have indicated that the usage of oxide electrodes instead of metals could improve the fatigue drastically, which is believed to be because the diffusion of oxygen from the electrodes fills the oxygen vacancies in the film. Furthermore, the oxygen in the film does not diffuse to the electrodes, which occurs in the case of metallic electrodes. The electrons of the metallic bonding may pin the switching domains in the film, so that the domain will be locked in this position and cannot contribute to the polarization.

The property of retention is the ability of the thin film to keep its polarization in time, which should be at least a few years. The imprint is a shifting of the hysteresis loop in the poling direction, which can lead to both read and write failure. Read failure occurs when the loss of the remnant polarization is too high, and write failure is when the shifting voltage of the hysteresis loop increases too much. These two different imprint failures are related to the operating voltage, and as it becomes smaller, the write failure will be the dominant problem. By the way of adding Sb or using oxide electrodes it is also possible to achieve the improved imprint characteristics. The imprint is also dependent on oxygen vacancies and ionic defects, like fatigue. Through Sb substitution to an Ti ion, the oxygen vacancy can be reduced through its higher valency.

Since the polarization ability of ferroelectric thin film is grain size dependent, the polarization decreases with decreasing grain size. Compatibility is a most important feature when depositing thin films. In the deposition process, one important feature is the lattice mismatch of the substrate and the thin film. If the mismatch is too big, it has been shown to be very difficult to grow the films. To solve this problem, buffer layers can usually be used to make the lattice mismatch smaller. Another compatibility problem arises when the ferroelectric material cannot handle the device manufacturing process. For example, hydrogen can be included in the IC device manufacturing process at high temperatures, which has a negative effect on the Bi in SBT. The film becomes Bideficient during the process by Bi diffusion from the grains to the grain boundaries. Pb is also volatile and can be damaging for the other materials in the device and/or degrading the desired composition ratios of the thin film, by diffusion during heat treatment. For memory application, the ferroelectric materials such as the perovskite structured Pb(Zr, Ti)O3 (PZT) and SrBiTaO (SBT) have been given most attention, in which the advantage of PZT is the larger polarizability, and of SBT, the better fatigue properties.

5.1.3. Environmental Aspects [12]

The impact of our modern lifestyle on the environment has moved from debates on the local environment to global environment issues. Recently, the EU had accepted a proposal of phasing out lead from all industries. This decision's major impact on the electronic industries is focused on the use of lead-containing solders in their products. But the law could be developed in a way that will force companies using Pb or other heavy metals to show that there exists no better material choice than Pb. In this way, companies can get a license to continue the use of Pb in their products. It might be difficult for a company to defend its device, if any other company is manufacturing the same device with other more environmentally friendly material choices. For example: PZT which contains Pb versus SBT which does not. In a longer perspective, SBT is a better choice as a material of research, although it does not have as good polarizability characteristics as PZT. Although the aim of today's technological physics activities is device application for their discoveries and research, they do not contain any notations concerning environmental aspects.

"Environment" mentioned in the technology field is related to an example such as lead's or bismuth's impact on the silicon environment, and there is research to understand different kinds of materials and their environmental characteristics. To achieve a long-term profitable project, one has to take into account the recycling as well as the desired characteristics of the product, when performing research aimed for commercial use. The use of PZT for memory applications is a short-term profitable project and it may not be of great importance in the future choice of material for memory applications or other devices by taking into consideration the environmental issues, as well as the problematic fatigue and process integration.

5.1.4. Pulsed Laser Deposition [13]

The PLD technique is a physical method for ferroelectric films deposition. It uses the laser pulse to make the material in the target so highly energized that the surface of the target explodes. This explosion throws out the evaporated material, perpendicularly, from the target, which is called the plume. The plume contains all components of the target and they are now in a very energized state. The purpose of the PLD process is to collect this plume upon a substrate, which is placed right in front of the plume, usually at a distance of a couple of centimeters from the target. The substrates temperature is a key parameter, especially when it comes to making high quality films with a preferred orientation. The deposition pressure and atmosphere also vary with the type of film being deposited.

The reach and shape of the plume can be controlled by changing either the deposition pressure or the fluence. The laser fluence is the energy per area of the laser spot hitting the target. It can be varied by either changing the laser pulse energy or by adjusting the laser spot size, which affects the plume's appearance both in reach and shape. The fluence has a minimum value which corresponds to the threshold fluence of the material. Below this threshold fluence, the energy is not high enough to create an explosion. A higher fluence will give a larger explosion which will make the plume become more round-shaped but not necessarily much longer. The other factor affecting the plume is the deposition pressure. When increasing the pressure, the plume will have more gas molecules to collide with and therefore it will not reach as far. When depositing PZT, the deposition pressure serves two purposes. The first one is to react with the plume and make sure that there is enough oxygen in the film to get the desired PZT phase. The second is keeping the Pb inside the film, since the Pb is very volatile, and the film would become Pb-deficient without the oxygen pressure.

The PLD technique has become a very attractive research tool, and the main parameters include: substrate temperature, laser fluence, target substrate distance, type of gas atmosphere (active or passive) and deposition pressure. However, industrialization of the PLD technique is still held back by the surface coverage since it is hard to apply to larger surfaces without affecting the homogeneity of the thin film.

5.1.5. Sol-gel Technique

The sol-gel technique has been used by the thin film coating industry for many years. This route involves the processes of hydrolysis and condensation for organometallic precursors on the substrate. The deposited sol is annealed to crystallize and densify the film. This process is advantageous considering its low temperature processing and its low cost. The starting solution contains metal organics of the desired ratio of the components. The solution is then either dripped, sprayed or by dipping applied to the surface. By drying the solution, most of the solvent is evaporated from the film to form a solid. After drying and firing, this solid is porous, making the film more densified. [14]

5.1.6. Pole Figure Plot

The pole figure plot is a stereographic projection showing the intensity of a known displacement in a certain direction corresponding to the surface of the sample. Performing a pole figure plot of the sample, will give all the necessary information to decide the orientations of the crystals.

The 2θ scan only provides information of the planes in parallel with the sample surface. This means that it does not give any information of the planes that are orientated away from parallel to the surface. The 2Φ scan can provide some information about this, but often it is not enough to fully characterize the orientation of the crystals in the sample. In the pole figure plot measurement, the 2θ angle is fixed, and the pole figure attachment is tilting and rotating the sample while measuring the intensities in every tilting and rotating angle (*alpha, beta*). The measuring data is then interpolated to make the pole figure plot.

The detailed mechanical process of the attachment while obtaining the data for the *pole figure plot*, is as follows. The 2θ angle is fixed and the sample is tilted *alpha*, around the θ–θ axis. Thus, it always keeps the angle of θ–θ towards the sample surface. For each *alpha*, the sample is rotated *beta* around the axis of the normal to the surface of the sample. Thus, it covers all the possible directions of the reference sphere. To make a complete *pole figure plot* that is to make it from *alpha* = 0 to 90°, it is necessary to perform the transmission and the reflection measurement.

5.1.7. Ferroelectric Thin Films

The piezoelectric thin film materials offer advantages in micro-electromechanical systems (MEMS), due to the large generated motions, with low hysteresis, high available energy densities, as well as high sensitivity sensors with wide dynamic ranges, and low power requirements. [15-19]

In MEMS applications, maximizing the piezoelectric coefficient is of considerable importance to reduce the driving voltage or increase the speed or sensitivity of a device. In a piezoelectrically actuated switch for rf signals, for example, the overall speed of the device is limited by the device size, with smaller dimensions yielding higher resonance frequencies. However, a certain maintained standoff is imperative, in order to maintain a good "on–off" ratio at high frequencies; while the device must be laterally small, it should still be possible for a vertical gap (ideally of several microns) to be maintained. If the actuator piezoelectric coefficient is large, this can be achieved at modest driving voltages.

Ferroelectric compositions are extremely attractive for the applications, where large thin film piezoelectric coefficients are required. But this comes at a cost for the complexity of the actuator composition. To solve this problem there is a need to develop film deposition routes that maintain stoichiometry, phase, and in some cases, orientation, and challenges associated with bringing new materials sets into cleanrooms. Fortunately, the piezoelectric MEMS community can benefit from the tremendous infrastructure development resulting from ferroelectric non-volatile access memories, which should speed implementation of ferroelectrics-based MEMS devices. The piezoelectric properties of ferroelectric thin films, the importance of morphotropic phase boundaries and high transition temperatures for MEMS applications will be emphasized.

Ferroelectric materials contribute to the available piezoelectric response, which would intrinsically be made due to the application of an electric field to an appropriately averaged ensemble of materials. The extrinsic response is typically associated with motion of non-180° domain walls or of phase boundaries. In thin films, the extrinsic contribution to the piezoelectric coefficient is considerably reduced in comparison with bulk materials, which limits the magnitude of the piezoelectric coefficient that is available in randomly oriented films. However, it has been demonstrated that the intrinsic piezoelectric coefficients are dependent on crystallographic orientation. In particular, many rhombohedrally-distorted perovskites, when poled and driven along the crystallographic [001], have extremely high piezoelectric coefficients. In contrast, in some of the tetragonal perovskites, the piezoelectric response is maximized when measured along the polar axis. Orientation can be controlled by deposition conditions, heating rate, and substrate choice, among other factors, which will enhance the piezoelectric properties.

Morphotropic Phase Boundaries [20-29]

Lead zirconate titanate (PZT) is the most widely utilized ferroelectric thin film for piezoelectric applications, which illustrates many of the important features of the other ferroelectric compositions. The lead zirconate titanate phase diagram is dominated by a rhombohdrally distorted ferroelectric region at low Zr contents, a tetragonal region at high Ti concentrations, separated by a morphotropic phase boundary (MPB), with a sliver of monoclinically distorted perovskite at lower temperatures near the MPB. In bulk ceramics and thin films, the maxima piezoelectric d and e coefficients are often observed at the MPB.

The differences in the correlation between factors like deposition method, film orientation, or film thickness, and enhanced properties in MPB, can be ascribed to the effect of grain size, or the relative poling efficiency in films of different compositions.

It is important for a good piezoelectric response system to display a morphotropic phase boundary with a high transition temperature. The magnitude of the piezoelectric coefficient can be enhanced if a system with a low transition temperature is utilized. However, this results in the expense of heightened temperature dependence in the piezoelectric response, increased tendency towards depoling, and often limitations on the available drive levels due to low coercive fields. Consequently, it is interesting to consider systems that show morphotropic phase boundaries with high transition temperatures. In the lead-based perovskites, the highest transition temperatures at known morphotropic phase boundaries occur in the PZT and $PbYb_{1/2}Nb_{1/2}O_3$-$PbTiO_3$ systems. In both cases, the transition temperatures are near 360 ^{o}C at the MPB. The higher transition temperature MPB compositions have been reported recently in $BiMeO_3$-$PbTiO_3$ solid solutions. For instance, high piezoelectric coefficients ($d33$ = 465 pC/N) have been measured in bulk ceramic $BiScO_3$-$PbTiO_3$ (36/64) with transition temperatures of 450◦C. Although the transition temperatures of thin films differ from those of bulk ceramic materials, for the large film thicknesses required in many MEMS devices, the transition temperatures are often reasonably close to those observed in bulk ceramics and single crystals. Therefore, the high Tc MPB ceramics offer a good guide as to the materials that will be interesting for high Tc, high piezoelectric activity piezoelectric thin films.

Effect of Film Orientation [30-31]

There is a strong orientation dependence of the piezoelectric properties in ferroelectric thin films. It is possible that the property differences are related to changes in the film domain state, perhaps due to residual stresses. The piezoelectric response also correlates with the magnitude of the remanent polarization; films with low remanent polarizations have low piezoelectric coefficients, which also indicate the importance of grain size in ferroelectric thin films, especially for films under tensile stress. In many thin films, the domain state is governed substantially by the stress, generated during the cooling process through the phase transformation temperature. Consequently, when exposed to tensile stresses during this cooling process, perovskite structured thin films tend to have their polar vectors in orientations approximately parallel to the substrate, and films under compressive stresses tend to have the polarization oriented more nearly perpendicular to the substrate. Since the mobility of non-180° domain walls in ferroelectric thin films tends to be limited compared to bulk materials, once the ferroelastic domain orientation is established, it is difficult to change at room temperature. Larger grained films tend to show more complicated domain structures within a single grain, resulting to be less susceptible to reduction in the remanent polarization and piezoelectric response.

For the example of PZT film, the processing offers the possibility to achieve textured films. Growth of PZT is strongly nucleation controlled, which can be manipulated to achieve a desired orientation. On a perovskite single crystal substrate or electrode, PZT grows almost inevitably in the same orientation as the template.

5.2. Development of PYN-PT Films

(1-x)Pb($Yb_{1/2}Nb_{1/2}$)O_3–$x$$PbTiO_3$ (PYbN–PT), one of the known relaxor ferroelectric–$PbTiO_3$ solid solutions, has the highest transition temperature (360 °C) near the morphotropic phase boundary (x: 0.5). The PYbN–PT bulk ceramics at compositions in the range from x= 0.2 to 0.49 show remanent polarizations as high as 32.8 μC/cm^2. Bornand *et al.* had reported the growth and ferroelectric properties of PYbN–PT epitaxial films. We have investigated the piezoelectric properties of PYbN–PT epitaxial films. Since the piezoelectric properties of relaxor-PT single crystals depend on the orientation and composition, in this article the piezoelectric properties of epitaxial (1–x)PYbN–xPT (x: 0.4, 0.5) films with (001) and (111) orientations were characterized.

By the pulsed laser deposition method, PYbN–PT films and $SrRuO_3$ bottom electrodes were prepared on (100) $LaAlO_3$, (100) $SrTiO_3$, and (111) $SrTiO_3$ single crystal substrates, and using sintered ceramics of PYbN–PT with (50/50) and (60/40) compositions, including 25 wt % excess PbO to compensate for lead loss during growth. Oxygen and ozone gases were introduced to the deposition chamber; after deposition, the films were quickly cooled in a 200 Torr oxygen/ozone atmosphere to room temperature. [32]

A. Phases and Structure

The growth for PYbN–PT epitaxial films was performed on (100) $SrRuO_3$ /(100) $LaAlO_3$ substrates using the (50/50) composition target. In the PYbN–PT epitaxial films, the pyrochlore phase was observed. Generally, in the deposition of Pb based perovskite films, the formation of a pyrochlore phase is mainly caused by either low growth temperatures or lead deficiency. The XRD shows the phase of the film changes from pyrochlore to perovskite with an increase in the SrRuO3 annealing time. A perovskite PYbN–PT film with high (001) orientation was obtained after vacuum annealing the substrate for 40 min. However, no significant difference in the XRD profile for the SrRuO3 itself was observed before and after the annealing. It is possible that the differences were associated either with surface reconstruction or volatility of ruthenium oxide. Consequently, subsequent growths were performed on annealed $SrRuO_3$ electrodes.

To optimize the growth conditions of the PYbN–PT films, various substrate temperatures and laser frequencies were exploited. The deposition pressure and the distance between the target and the substrate were fixed at 400 mTorr and 45 mm. Before deposition ferroelectric films, (100) $SrRuO_3$ /(100) $LaAlO_3$ substrates were annealed at 790 °C for 40 min. The growth rates at laser frequencies of 3 and 5 Hz were 60 and 90 nm/min, respectively.

From the XRD curve and the rocking curve, the quantity of pyrochlore phase was estimated from the ratio of the diffraction peak intensities of the pyrochlore phase. At 670 °C andat laser frequencies of 3 Hz , the quantity of pyrochlore phase increased to 6%. In contrast, (001) PYbN–PT films with 0.5% pyrochlore phase could be achieved for temperatures below 680 °C at a 5 Hz pulse rate. Thus, higher growth rates are required to obtain perovskite PYbN–PT at higher temperatures. These results are consistent with increased Pb volatilization at higher temperatures favoring pyrochlore phase formation. For higher laser pulse rates, the delivery rate for PbO was sufficiently high that the perovskite phase could be stabilized to a higher temperature.

Using the growth condition of a laser repetition rate of 3 Hz and 650–660 °C, PYbN–PT (50/50) films were deposited on (100) $SrRuO_3$ /(100) $SrTiO_3$ and (111) $SrRuO_3$ /(111) $SrTiO_3$ substrates. Since the piezoelectric properties of films may depend on the substrate due to differences in the thermal expansion coefficient, both (100) $SrRuO_3$ /(100) $LaAlO_3$ and $SrRuO_3$ /(100) $SrTiO_3$ was used.

The optimum growth temperature depended on the substrates. PYbN–PT films with good phase purity were obtained on (100) $SrRuO_3$ /(100) $SrTiO_3$ substrates. However, PYbN–PT films deposited on (111) $SrRuO_3$ /(111) $SrTiO_3$ substrates were not well crystallized.

B. Ferroelectric/Dielectric/Piezoelectric Properties

PYbN–PT (50/50) and (60/40) epitaxial films exhibit well-saturated hysteresis loops. In both these two compositions, (001) epitaxial films showed lower dielectric constants and higher remanent polarizations than (111) Epitaxial films. In (111) epitaxial films, the polarization in some regions of the films is largely in-plane and is not switched by the applied electric field due to limited non-180° domain wall motion. In both the epitaxial films, the (50/50) films have larger remanent polarization than the (60/40) films.

The temperature dependence of the dielectric constant and loss for the PYbN–PT epitaxial films were measured during heating. For (50/50) and (60/40) compositions, the dielectric constant has a maximum near 380 and 320 °C. The maximum dielectric constant of the (50/50) film (3200) was higher than that of the (60/40) film (1900). The behavior of the dielectric constant is consistent with the data obtained in bulk ceramics with the same composition. These results suggested that the *B*-site stoichiometry of both the films was nearly that of the targets. The relatively high dielectric losses (5%–7%) are likely due to the domain wall motion, since the dielectric losses decreased to 2% by applying a dc bias of 20 V.

To measure the piezoelectric coefficients, PYbN–PT films were poled with a dc electric field at room temperature. From the variation of the $e_{31,f}$ coefficient for the (001) PYbN–PT (50/50) film on (100) $SrRuO_3$ /(100) $LaAlO_3$ as functions of the poling electric field and time, there was significant asymmetry in the $e_{31,f}$ coefficient as a function of the poling direction. The asymmetric behavior of the $e31, f$ coefficient can be explained by the internal bias electric field of the films, since the hysteresis loops were shifted along the electric field axis. Similar asymmetric behavior of the $e_{31,f}$ coefficient was observed in other PYbN–PT films, when the hysteresis loops were shifted towards the positive direction.

The epitaxial growth and transverse piezoelectric properties of $(1\text{-}x)Pb(Yb_{1/2}Nb_{1/2})O_3$-$xPbTiO_3$ films deposited by pulsed laser deposition was reported. The surface condition of the $SrRuO_3$ bottom electrodes, as well as the growth temperature and lead deficiency, affected the phase formation of PYbN–PT. After vacuum annealing the $SrRuO_3$ bottom electrodes, (001) PYbN–PT perovskite thin films were fabricated for a wide range of deposition rates (60–90 nm/min and temperatures ~620–680 °C). Both (50/50) and (60/40) compositions (001) PYbN–PT epitaxial films show good crystalline quality. The (001) films exhibited higher remanent polarizations and piezoelectric $e_{31,f}$ coefficients than the (111) films. The transition temperatures of the (50/50) and (60/40) films were near 380 and 320 °C.

5.3. Development of BSPT Thin Films

Recently, $BiScO_3$-$PbTiO_3$ has been reported as a new piezoelectric with a high T_c. At the morphotropic phase boundary(MPB) composition, T_c of $BiScO_3$-$PbTiO_3$(64/36) is 450°C, which is higher than that of Pb(Zr, Ti)O_3. A remanent polarization of ~32μC/cm^2 and a d_{33} piezoelectric coefficient of ~465pC/N were observed in bulk ceramics. BSPT epitaxial films were investigated in terms of their ferroelectric and piezoelectric properties and the recent progresses of these materials are overviewed. [33-47]

Today, there exist several physical and chemical methods for preparing thin films. The physical methods can be divided into thermal evaporation and sputtering and the chemical into a vapor phase and liquid phase. Many methods were used to prepare the BSPT thin films in order to get better ferroelectric and piezoelectric properties.

The epitaxial BSPT thin films with high Curie temperature were prepared via PLD, sol-gel method and hydrothermal method on the different substrates. The remnant polarization and coercive electrical field were similar to the behavior of BSPT single crystal. The BSPT thin films were investigated about the mechanical properties, graded thin films and orientation of the sol–gel-derived BSPT thin films. A mechanism for the orientation evolution in the sol–gel-derived BSPT thin films was proposed. The compositionally graded thin films could present higher properties than the homogeneous thin films and might be attractive candidates for many applications. However, the high cost of Sc precludes the use of BSPT in some applications.

The main properties of several BSPT thin films are shown in Table 1. [33-47]

5.3.1. $BiScO_3$–$PbTiO_3$ Epitaxial Films by PLD

Epitaxial films of (001) $BiScO_3$– $PbTiO_3$ were grown by pulsed laser deposition using ceramic targets of (1-x)$BiScO_3$– $x$$PbTiO_3$ (x:0.5) at substrate temperatures of 700 °C on (100) $SrRuO_3$ /(100)$LaAlO_3$ substrates, and the ferroelectric and piezoelectric properties were investigated. The remanent polarization was 36 μC/cm^2 and well-saturated hysteresis loops were observed. The films had a room temperature dielectric constant of 800, tan δ of 0.09, a maximum permittivity of 3600 at 460 °C, and the room temperature e_{31f} piezoelectric coefficient of 29 C/m^2. The piezoelectric properties obtained were comparable to those of Pb(Zr, Ti)O_3 or relaxor ferroelectric-$PbTiO_3$ films. [40]

Recently, $BiScO_3$– $PbTiO_3$ has become a new piezoelectric with a high Tc of 450 oC near the MPB composition. A remanent polarization of 32 mC/cm^2 and a d_{33} piezoelectric coefficient of 465 pC/N were observed in bulk ceramics. Since rhombohedrally distorted perovskite ferroelectrics show higher piezoelectric properties along the (001) direction, (001) $BiScO_3$– $PbTiO_3$ epitaxial films in the rhombohedral phase were prepared.

$BiScO_3$–$PbTiO_3$ films were deposited using sintered ceramics of (1-x)$BiScO_3$– $x$$PbTiO_3$ (x:0.5) which were batched with 20 mol% excess PbO and 10 mol% excess Bi_2O_3 to compensate for the loss of lead and bismuth during growth. In XRD of $BiScO_3$– $PbTiO_3$ films, although small peaks corresponding to polycrystalline $BiScO_3$–$PbTiO_3$ were observed, no second phase such as pyrochlore phase was present. The out-of-plane lattice constant determined by using the Nelson–Riley method was 4.0516 Å, which is consistent with the lattice constants reported in bulk ceramics of rhombohedral $BiScO_3$–$PbTiO_3$. The full width at half maximum (FWHM) of the 002 diffraction peaks were 0.29° and 1.08°, respectively.

Table 5.1. Main properties of BSPT thin films

material	method	substrates	T_c	thickness	P_r	dielectric constant	d_{33}^*	tan δ	permittivity	e_{31f}
BSPT50	PLD	(100) $SrRuO_3$ /(100)$LaAlO_3$	460°C	1 μm	36 $\mu C/cm^2$	800		0.09	3600	-9 C/m^2
BSPT66	Sol-gel	Pt(111)/Ti/SiO_2/Si	435°C	350nm	74 $\mu C/cm^2$					
(100)-oriented PSPT66	Modified Sol-gel	Pt(111)/Ti/SiO2/Si(111)	435°C		33 $\mu C/cm^2$	1200	~70pm/V	0.05		
BSPT66	Sol-gel	(100) (Nb-STO)			74 $\mu C/cm^2$		130pm/V			
BSPT	hydrothermal	Nb-doped $SrTiO_3$ [100] single crystal		3.1μm	30 $\mu C/cm^2$		65pm/V			

Pt top electrodes were sputtered through a shadow mask, to characterize the electrical properties. Polarization hysteresis loops were measured by a Radiant Technologies RT66A standard ferroelectrics tester. The dielectric constant and loss were determined using an inductance-capacitance resistance meter (HP4284A).

From the observed well-saturated hysteresis loops, the remanent polarization and coercive field were 36 $\mu C/cm^2$ and 65 kV/cm. The dielectric constant was 800 at room temperature and exhibited a maximum near 460 °C. The temperature of phase transition from rhombohedral to tetragonal distortions was 300 °C. It is encouraging that the dielectric loss remains below 0.1 up to 480 °C.

For the measurements of the e_{31f} coefficients, a dc electric field was applied to the top electrode of the films at room temperature for poling. The e_{31f} coefficient after poling with a positive electric field was almost constant with the poling electric field and time. The value obtained (29 C/m^2) was comparable to that of Pb(Zr, Ti)O_3 or relaxor ferroelectric-$PbTiO_3$ films.

5.3.2. $BiScO_3$–$PbTiO_3$ Thin Films Synthesized by Sol-Gel Method

100-oriented $0.34BiScO_3$–$0.66PbTiO_3$ thin films were synthesized on Pt111 /Ti/SiO_2/Si substrates using a sol-gel method. From the saturated polarization hysteresis loops, a remanent polarization of 33 C/cm^2 was achieved. The dielectric constant and loss for the thin films were 1200 and 5%, with the Curie temperature of 435 °C, over 80 °C higher than that of the PbZr, TiO_3 (PZT) thin films. The local effective piezoelectric coefficient d_{33*} was approximately 70 pm/V. [35]

Various techniques have been employed to fabricate the ferroelectric thin films, among which the sol-gel method has obtained particular interest due to its advantages of chemical homogeneity, facility of stoichiometry control, low cost, and easy integrability into semiconductor devices. Extensive studies have been conducted on fabricating oriented thin films via the sol-gel method, such as introducing a seeding layer or using single crystal substrates.

$0.34BiScO_3$–$0.66PbTiO_3$ precursor solution with 18% lead excess was prepared by using lead acetate, tetrabutyl titanate, scandium acetate, and bismuth nitrate as starting materials and glacial acetic acid, isopropyl alcohol, and distilled water as solvents. Formamide and acetylacetone were added to avoid the appearance of cracks. The resultant solution, with a concentration of 0.24 M, was aged for 10 days before deposition. The as-deposited organic layer was spin-coated on Pt /Ti/SiO_2/Si substrates at 5000 rpm for 30 seconds, dried at 80 °C for 1 hour, and pyrolyzed at 350 °C for 3 minutes. This process was repeated many times to achieve a desired thickness. Additional annealing treatment was carried out at 700 °C in an oxygen atmosphere using a rapid thermal processing RTP furnace.

For the structure study, a single perovskite phase is observed and no second phase such as a pyrochlore phase is present the BSPT samples. The intensity maximum appears at the 100 peak, while only a weak 110 peak and no 111 peak have been detected. A Lotgering factor used to describe the level of orientation was 0.81 for the BSPT thin film.

Commonly, Pb-based ferroelectric films directly deposited on the Pt111 /Ti/SiO_2/Si111 substrates show 111 preferential orientation, due to the lattice match. The 100 orientation in the BSPT thin films has been attributed to three factors: RTP annealing process; the 18% excess lead addition to the precursor solution and the drying process at 80 °C for 60 min. However, the reason that the long drying process results in 100 orientation is still unclear.

For studying the electric properties, Pt top electrodes with a diameter of 0.2 mm were sputtered. The achieved saturated hysteresis loops indicated the remanent polarization and coercive field are 33 C/cm^2 and 220 kV/cm. The remanent polarization is comparable to the reported value of the BSPT films by pulsed laser deposition while the coercive field is much higher.

The reason may be the sol-gel derived BSPT thin film has much smaller grain size 40 nm, than the pulsed laser deposited one (250 nm). To evaluate the T_C, the temperature dependence of the dielectric constant and loss for the BSPT thin film were measured at 100 kHz, in which room dielectric constant and dielectric loss are approximately 1200 and 5%. A dielectric constant peak of 6300 appears near 435 °C (Tc), which is 25 °C lower than that observed in bulk BSPT ceramics.

The surface morphology and local effective piezoelectric coefficients d_{33} for the BSPT thin film were characterized using the SPM method. The topography for the as-prepared BSPT thin film, measured by SPM using an atomic force microscopy AFM mode showed the grain size of approximately 40 nm, and the measured roughness rms is 6.4 nm. The measurement of the local effective piezoelectric coefficients d_{33} was achieved by keeping the SPM tip fixed above the interesting grain and applying a dc voltage from −9 to 9 V while recording the piezoresponse signal. The local effective d_{33} dependences of the applied voltage show well-shaped piezoelectric hysteresis loops, from which the effective d_{33} values are estimated to be about 70 pm/V. It is encouraging that the sol-gel derived BSPT with a 100 orientation degree of 80% shows comparable piezoelectric response to the highly 100-oriented PZT thin films 80 pm/V, which suggested such sol-gel derived BSPT thin film potentially interesting for electric devices employed at high temperature.

Effect of Crystallization Process [47]

Two different kinds of crystallization processes were adopted to prepare sol-gel derived $BiScO_3$–$PbTiO_3$ thin films. One is the single crystallization, the other, is multiple crystallization. The effect of different crystallization processes of single crystallization and multiple crystallization on the ferroelectric properties of sol-gel derived BSPT thin films were addressed. The films prepared in the first process had a higher remanent polarization and dielectric constant. The remanent polarization and dielectric constant were 50 μC/cm^2 and 1609 for the single crystallization sample, and 40 C/cm^2 and 1429 for the multiple crystallization one. The Rayleigh law study indicated that the multiple crystallization process generated more irreversible domain wall motion for the ferroelectric films, with respect to the single crystallization process, which is attributed to the layer-to-layer interfaces formed during the multiple crystallization processing, acting as an energy barrier, pinning the domain wall motion, and resulting in more irreversible domain wall motion.

The precursor solution with a composition of $0.34BiScO_3$–$0.66PbTiO_3$ with 5 mol % lead excess was synthesized via an aqueous sol-gel method. The tetragonal BSPT thin films were deposited onto Pt111 /Ti/SiO_2/Si111 substrates by a spin coating method. The wet layers were pyrolyzed and crystallized at 450 °C for 5 min and 700 °C for 6 min in ambient atmosphere, by using a rapid thermal processing (RTP furnace, the sample prepared by a multiple crystallization process was named BSPT-a, the other one, BSPT-b). Both the samples show pure perovskite phases, without any second phases such as pyrochlore. The FE-SEM observation on the microstructures reveals that the two films have a similar grain size of approximately 40 nm and film thickness of 250 nm.

From the ferroelectric hysteresis loops measured at 100 Hz for the two thin films, both of the two samples show saturated hysteresis loops. BSPT-b shows a larger polarization value of 56 μC/cm^2, in comparison with the remanent polarization of 40 μC/cm^2 for BSPT-a. Moreover, BSPT-b exhibits larger dielectric constant than BSPT-a (which are 1609 and 1429, respectively), but with the same dielectric losses of 0.08. The dc bias dependence of the dielectric constant for the two samples, were measured under a small oscillation signal of 0.5 V at 1 kHz from which we found that the dc bias has a more significant effect on the dielectric constant of BSPT-a than that of BSPT-b, since the dielectric constant of BSPT-a exhibits a higher increase in the small dc bias field region and a larger decrease in the large dc bias field region, since the influence of dc bias field on the dielectric properties is mainly due to the pinning effect of dc bias field on the domain wall mobility. This implies that the domain wall motion is easier to be pinned by the dc bias field in BSPT-a than in BSPT-b.

There are two kinds of domain wall motion contributions to the polarization and dielectric constant in ferroelectric materials. Those not being pinned by defects are reversible, while those interacting with defects and being pinned from moving are irreversible, which may decrease the polarizability.

In the future, more detailed investigations such as TEM observation needs to be carried out to reveal the contribution of layer-to-layer interfaces generated during the multiple crystallization process on the properties.

Mechanical Properties by Nanoindentation [36]

Both the electric properties and the mechanical properties of the ferroelectric thin films are very important for the design of the microdevices. The piezoelectric coefficients are dependent on mechanical coefficients, correlating to the mechanical state of the materials. For instance, the piezoelectric coefficient d_{33} of a ferroelectric thin film is actually an effective coefficient dependent on the mechanical coefficient due to the clamping effect of the substrate. Thus, when designing a piezoelectric actuator or sensor, it is important to learn the mechanical properties and states of the ferroelectric thin films. Moreover in MEMS design, the mechanical properties play critical roles in the delamination, brittle fracture, and fatigue degradation of the multilayer thin film structures, which cause poor reliability that is one of the major problems. Thus the investigation of the mechanical properties of the ferroelectric thin films is of prime importance for the practical applications.

Various methods can be used for measuring the mechanical properties of thin films, such as atomic force microscopy, Brillouin light scattering, and the impulse acoustic method. Among these methods, the nanoindentation technology is frequently used, due to the ease of sample preparation, the inherently simple estimation of the mechanical coefficients and the high resolution. Key mechanical properties such as hardness and elastic modulus can be determined without imaging the indentation impression from the load-displacement data during an indentation test.

Through nanoindentation methods, the mechanical properties of sol-gel derived $BiScO_3$–$PbTiO_3$ BSPT thin films were investigated. The effects of annealing temperature on the microstructures and mechanical properties were indicated. XRD, SPM, and nanoindentation with a diamond tip were used to investigate the microstructures as well as the mechanical properties of sol-gel derived $BiScO_3$–$PbTiO_3$ thin films annealed at different temperatures.

As the annealing temperature increased from 550 to 800 °C, the 100 orientation degree of the BSPT thin films increased, from a Lotgering factor of 0.47 to 0.67, with the grain size increased from 22 to 54 nm. The hardness for the BSPT thin films dropped from 7.8 to 6.7 GPa, due to the grain size effect. The Hall-Petch equation can be used to explain and fit the experiment results, with a probable lattice friction stress of 4.69 GPa and a Hall-Petch constant of 15.35 GPa nm/m^2 obtained for the BSPT thin films. Because of the combined effect of grain size and film orientation, Young's modulus did not show a clear trend with the annealing temperature, as the highest Young's modulus was 132 GPa.

Sol–Gel Derived Compositionally Graded $BiScO_3$–$PbTiO_3$ Thin Films [44]

The high performance compositionally graded ferroelectric thin films can be obtained on the basis of the sol–gel method. Multilayer thin films with compositional fluctuation across the thickness exhibit striking properties, which were not observed in conventional ferroelectric films. Compared with the single-layer films, the enhanced chemical heterogeneity of the compositionally graded films leads to a higher tunability; and the enhanced piezoelectric strain in the graded structure can be deduced by the built-in strain gradient that arose from the grading of the composition and the spontaneous self-strain.

The sol-gel derived compositionally graded $BiScO_3$–$PbTiO_3$ (BSPT) thin films could be fabricated on Pt/Ti/SiO_2/Si. Three types of thin films were prepared. (i) The homogeneous thin film had a homogeneous composition of $0.36BiScO_3$–$0.64PbTiO_3$, which was close to the MPB reported in the bulk BSPT ceramics. (ii) The up-graded thin film was deposited by the sequence of two layers for each composition from $0.45BiScO_3$–$0.55PbTiO_3$ at the substrate to $0.2BiScO_3$– $0.8PbTiO_3$ at the surface. (iii) The down-graded thin film was deposited by the sequence of two layers for each composition from $0.2BiScO_3$–$0.8PbTiO_3$ at the substrate to $0.45BiScO_3$– $0.55PbTiO_3$ at the surface. For the compositionally graded thin films, the thickness of each composition layer was approximately 50 nm.

In up-graded thin film, the $PbTiO_3$ content increased from the film–substrate interface to the surface of the film, the down-grade thin film showed the opposite trend. These compositionally graded BSPT thin films had similar remanent polarization value and a higher dielectric constant, dielectric tunability, and piezoelectric coefficient d_{33}, compared with the homogeneous thin film with a same composition of MPB BSPT films.

Therefore, the compositionally graded thin films could present higher properties than the homogeneous thin films, and might be attractive candidates for many applications.

Film Orientation Control [46]

The effects of different factors on the orientation of the sol–gel derived $BiScO_3$–$PbTiO_3$ (BSPT) thin films fabricated on Pt(111)/Ti/SiO_2/Si(111) substrates were investigated. The higher lead excess concentration, longer drying time, higher pyrolysis temperature, longer pyrolysis time, higher crystallization temperature, and longer crystallization time could enhance the (100) orientation of the BSPT thin films. The mechanism for the orientation evolution in the sol–gel derived BSPT thin films was proposed to be the attribution of the (100)-oriented PbO nanocrystals forming during the pyrolysis process due to the lattice match.

When the pyrolysis temperature is higher than 450 °C, PbO nanocrystals precipitate from the amorphous film, which occur at the film–substrate interface, and tend to be (100) oriented, acting as seeds for the nucleation of BSPT (100) due to lattice match. During the crystallization process, the heterogeneous nucleation takes place on these seeds, leading to the BSPT (100) orientation.

$BiScO_3$-PbTiO3 Ultrathin Film [48]

There are few reports on BSPT thin films of near-MPB composition. Pulsed laser deposition and sol-gel methods have been used to fabricate BSPT thin films. These reports indicate that BSPT thin films have good thermal stability and piezoelectric properties. However, the previous reports on BSPT thin films were concerned only with fully relaxed states, meaning that the thin films were thicker than 100 nm.

The ferroelectric/piezoelectric properties correlate with the structural properties of MPB (1-x)$BiScO_3$-(x)$PbTiO_3$ (BSPT, x:0.64) epitaxial thin films on $SrTiO_3$ (001) and Nb:$SrTiO_3$ (001) with various thicknesses that were studied. The films were deposited using off-axis radio frequency magnetron sputtering. The thicknesses of our thin films ranged from 8 to 60 nm. The strong thickness dependence of roughness of thin films can be observed in x-ray reflectivity data and demonstrates the importance of strain relaxation due to formation of misfit dislocations. Clear piezoresponses and polarization reversal phenomena can be observed in the ultrathin limit down to 8 nm through Piezo Force Microscope experiments. The piezoresponse data as a function of thickness correlates with the structural modification of thin films. These epitaxial BSPT thin films, with excellent piezoelectric properties, will be useful for next-generation piezoelectric applications.

5.3.3. Sol-Gel Derived $BiSc_{1/2}Fe_{1/2}O_3$–$PbTiO_3$ Piezoelectric Films [49]

Among the different raw materials, Sc is very expensive. A substituent for Sc_2O_3 by other materials can reduce the cost of the raw materials that are used. (1-x)BiSc1-yFeyO$_3$–x$PbTiO_3$ (BFS–PT) solid solutions were fabricated for achieving this aim, and for y = 0.5, BFS–PT ceramics with the MPB compositions (x = 0.5–0.55) showed a piezoelectric coefficients (d_{33})

of 300 pC/N. This was similar to that of hard PZT composition, but the curie temperature of this system is 440 °C.

These compositions of polycrystal BSPT thin films can be grown on platinized silicon substrates via the sol–gel method, epitaxial BSPT thin films were prepared on (1 0 0) $SrRuO_3$/(1 0 0) $LaAlO_3$ single crystal substrates via pulsed laser deposition, and subsequently on Nb-doped $SrTiO_3$ (1 0 0) single crystal substrates via the sol–gel method. Lead acetate [$Pb(COOCH_3)_2$], tetrabutyl titanate [$Ti(OC_4H_9)_4$], scandium acetate [$Sc(COOCH_3)_3$], bismuth nitrate [$Bi(NO_3)_3$], and iron nitrate [$Fe(NO_3)_3$] were chosen as the raw materials with glacial acetic acid, isopropyl alcohol and distilled water as the solvents.

Four solutions in the proportions of x =0.6, 0.55, 0.5, 0.45 with 10% lead excess were adopted to synthesize the thin films deposited on Pt(1 1 1)/Ti/SiO2/Si(1 1 1) substrates, then dried at 100 °C and pyrolyzed at 400 °C in ambient atmosphere. This process was repeated 10 times to obtain a desired thickness (400 nm) for the BFS–PT films. Subsequently these films were annealed at 700 °C.

The XRD patterns of the sol–gel derived BFS–PT thin films with different compositions show pure perovskite phases, without any second phases such as pyrochlore existing to the detection limits of our diffractometer. No coexistent peaks of tetragonal phase and rhombohedral phase have been found for all films, suggesting a solid-solution phase is formed at the composition range. The surface morphologies of these BFS–PT thin films with different compositions exhibit similar surface grain structures and grain size distributions, and the mean grain size of all these films are about 75 nm.

The ferroelectric hysteresis loops of all the BFS–PT films measured at 100 Hz were well saturated, with high direct current (DC) resistivity, which indicates few defects existing in these thin films. The highest remanent polarization (Pr) is obtained at 35 $\mu C/cm^2$ for x = 0.55, and the coercive field (Ec) values of about 110 kV/cm for all these films are similar.

The DC bias dependence of the dielectric constant for these BFS–PT thin films shows typical ferroelectric butterfly loops for all the samples. The dielectric loss of these thin films, present a similar value of about 0.05. For ferroelectric films, the contributions to the permittivity can be divided into intrinsic and extrinsic components. The extrinsic contribution which mainly originated from the domain wall motion will be restricted by the direct current bias field when the field is larger than the coercive field. The permittivity values for the BFS–PT films with different compositions become more and more similar as the DC bias field increases. The intrinsic permittivity exhibits a weak relationship with the composition and the enhanced dielectric properties at MPB composition may be attributed to the extrinsic contribution which is mainly originated from the reversible domain wall motion.

The BFS–PT thin films also exhibit typical ''butterfly'' shape displacement-voltage loops in piezoelectric property studies, which present the largest displacement for MPB composition samples (x = 0.55), and the effective piezoelectric coefficient d_{33} can be calculated from these loops, which are about 42 pm/V, 50 pm/V and 40 pm/V for x = 0.6, x = 0.55, x = 0.5, respectively. These piezoelectric properties are consistent with the dielectric results, indicating that the enhancement of the piezoelectric activity around MPB composition may also be attributed to the larger activity of domain wall motion.

5.3.4. Growth for $BiInO_3$–$PbTiO_3$ Films [50]

The chemical solution deposited $x$$BiInO_3$–1−$x$$PbTiO_3$ (x: 0.10-0.35) thin films on platinized silicon substrates have been fabricated, using a $PbTiO_3$ seed layer. The dielectric, ferroelectric, and piezoelectric properties pure phase $x$$BiInO_3$–1−$x$$PbTiO_3$ thin films were also investigated. 0.15$BiInO_3$–0.85$PbTiO_3$ film with a 470 nm thickness, exhibit the room temperature permittivity of 650, while the dielectric loss tangent was below 2%. The coercive field and remanent polarization were 73 kV/cm and 22 μC/cm^2. The ferroelectric transition temperatures of the $x$$BiInO_3$–1−$x$$PbTiO_3$ x=0.10–0.20 films were all in excess of 550 °C. For x=0.15, the $e31,f$ piezoelectric coefficient was −2.7 μC/m^2.

Eitel *et al.* showed that MPB compositions of Bi*Me*O_3–$PbTiO_3$ (*Me* =Sc, In, and Yb) with reduced tolerance factors have higher T_C than that of PZT (350 °C). For example, MPB $BiScO_3$–$PbTiO_3$ BS-PT ceramics have a T_C=450 °C and d_{33}=460 pC/N piezoelectric coefficient. Unfortunately, the high cost of Sc precludes use of this material in some applications. $BiInO_3$–$PbTiO_3$ (BI-PT) and $BiYbO_3$–$PbTiO_3$ (BYb-PT) binary systems have also the high T_C of 500 °C. It is difficult to obtain pure phase perovskite samples due to their intrinsic instability in the perovskite structure; and of these compositions, thin film synthesis and properties have been reported only in the BS-PT system. Epitaxial 0.5$BiScO_3$–0.5$PbTiO_3$ and 0.4$BiScO_3$–0.6$PbTiO_3$ films were grown by pulsed laser deposition, and showed transverse piezoelectric coefficients of $e31,f$–9 to −12 C/m2. Subsequently, sol-gel derived BS-PT films were reported. However, there is no literature on thin film $BiInO_3$ or $BiYbO_3$-based solid solutions. The processing, as well as the dielectric, ferroelectric, and piezoelectric properties of BI-PT thin films fabricated by chemical solution deposition CSD are investigated.

Bismuth nitrate pentahydrate, indium nitrate hydrate, lead acetate trihydrate, and titanium butoxide were used as raw materials to prepare the precursors. The BI-PT precusor solution preparation procedures were done at room temperature, and the molarity of the final solution was around 0.2*M*.

Direct deposition of the films on Pt/SiO_2 / Si substrates may result in randomly oriented and mixed phase films. Therefore, to mitigate this problem, a $PbTiO_3$ seed layer was deposited before the BI-PT films to provide a perovskite template, as well as to increase the degree of 100 film orientation. For this purpose, a 0.15*M* $PbTiO_3$ solution with 40 mol % PbO excess was spin-coated on the substrate and then pyrolyzed at 250 °C for 1 min and 400 °C for 1 min on a hot plate. The sol-gel derived $PbTiO_3$ layer with a thickness of 20 nm was crystallized at 580 °C for 1 min with oxygen flow in a rapid thermal annealer. Following crystallization of this layer, the BI-PT solution was spun onto the substrate. The films were then pyrolyzed sequentially and annealed in oxygen. These procedures were repeated until the desired film thicknesses of 470 nm with 8 layers were obtained.

From the x-ray diffraction XRD patterns, the calculated lattice parameters of $x$$BiInO_3$–1−$x$$PbTiO_3$ films as a function of composition were indicated. All of these films were annealed at 600 °C. The $x$$BiInO_3$–1−$x$$PbTiO_3$ films were phase pure for 0.10-0.25. An In_2O_3 second phase was observed for films with x=0.3. However, when the annealing temperature was reduced to 550 °C, the In_2O_3 secondary phase disappeared from the XRD patterns. Moreover, it was found that once the perovskite phase formed, it was retained without decomposition at higher temperatures 700 °C XRD patterns are not shown here. Using this approach, pure perovskite BI-PT films were achieved up to x=0.35. This suggests that the

$PbTiO_3$ buffer layer expands the composition range of perovskite phase stability to a larger concentration of $BiInO_3$. In bulk ceramics, the limit of solid solution of BI-PT was at x=0.25 for sol-gel derived powders.

From the XRD patterns, as the $BiInO_3$ content increases, a reduction in the perovskite phase peak intensities and a broadening of the XRD peaks occurs. The film peaks were then fit using *JADE software* to determine the lattice parameters. In order to separate the contributions of the $PbTiO_3$ seed layer from the film itself, each family of peaks was fit using two separate peaks. The smaller lattice parameter was almost constant 3.94 Å over the range of compositions. The measured lattice parameter of a chemical solution deposited $PbTiO_3$ film without any $BiInO_3$–$PbTiO_3$ on top was 3.9280 Å. When a BI-PT x=0.30 film was pyrolyzed on a $PbTiO_3$ buffer layer without a post annealing step, then the lattice parameter increased to 3.938 Å, which is identical to the smaller of the two lattice parameters of the post annealed BI-PT x=0.30 film. This suggests that there is some incorporation of $BiInO_3$ into the $PbTiO_3$ buffer layer. Thus, this constant lattice parameter was ascribed to the template layer.

The length of the two lattice parameters decreased from 10 mol % BI to 15 mol % BI and then linearly increased up to 35 mol % BI, which would correspond to the loss of the tetragonal crystal structure near 15 mol % $BiInO_3$. This is consistent with the results of BI-PT ceramics. The linear increase suggests that the BI-PT films form a solid solution over this composition range based on Vegard's law. The error range of lattice parameters increases as BiInO3 content increases because of broadened peaks.

Pt was sputtered as a top electrode on the films through a metal shadow mask, to investigate the dielectric, ferroelectric, and piezoelectric properties. The polarization-field hysteresis loops $0.15BiInO_3$–$0.85PbTiO_3$ film annealed at 600 °C are well-shaped, and the coercive field and remanent polarization were 73 kV/cm and 22 C/cm^2.

The temperature dependence of the dielectric constant and dielectric loss for three different compositions were obtained. For $0.15BiInO_3$–$0.85PbTiO_3$, the dielectric constant was 650 at 10 kHz at room temperature, and only weakly dependent on frequency. The room temperature loss tangent was below 2%. As the BiInO3 content increased, the temperature of the maximum dielectric constant increases up to 582 °C. Combined with the low loss tangent to elevated temperatures, this suggests that BI-PT films could expand the usage temperature range of piezoelectric films up to 300 °C. No clear signatures of a lower temperature ferroelectric-ferroelectric phase transition are apparent for any of the compositions shown.

By the wafer flexure method, the $e_{31,f}$ coefficients were measured. For poling prior to measurement, a dc bias was applied to the top electrode at room temperature. The $0.15BiInO_3$–$0.85PbTiO_3$ film poled at $3Ec$ for 20 min exhibits the $e_{31,f}$ coefficient of −2.7 C/m^2, which is lower than those of BS-PT and PZT family films. The piezoelectric properties are also strongly composition dependent. Density-functional theory calculations suggested that the MPB of BI-PT would be at 0.33BI-0.67PT. In contrast, other experimental work suggests that the MPB would appear somewhere between 5%–15% BI. If the calculations are correct, then the comparatively low $e_{31,f}$ value for the film with x=0.15 could result in part from being far from the MPB. In any case, this work provides a set of baseline measurements for phase-pure BI-PT.

In conclusion, BI-PT films were grown by CSD with good perovskite phase stability. The combination of a reasonable $e_{31,f}$ of −2.7 C/m^2 and a high T_C of 581 °C for the $0.15BiInO_3$–$0.85PbTiO_3$ films encourage future exploration on the use of BI-PT films for sensors and actuators in harsh conditions.

5.3.5. High-Tc Nano Ferroelectrics

$BiScO_3$–$PbTiO_3$ Powders and Thin Films Using Metal-organic Precursor Solutions [51]

The perovskite BS-PT powders and thin films near their MPB composition, with various $PbTiO_3$ concentrations, were prepared using a chemical process based on metalorganic precursor solutions. This chemical solution process is useful for acquiring high homogeneity, flexible control of the chemical composition, and for lower processing temperatures to reduce environmental load. The crystallization and phase transition of chemically synthesized BS-PT powders and thin films, and dielectric and ferroelectric properties of the BS-PT films on Pt/TiOx/SiO2/Si substrates were investigated.

$Bi(OtC5H11)_3$, $Sc(OiC_3H_7)_3$, $Pb(CH_3COO)_2$, and $Ti(OiC_3H_7)_4$ were selected as starting materials for preparing $(1\text{-}x)BiScO_3\text{-}xPbTiO_3$ (x = 0.50, 0.60, 0.65, 0.70) precursor solutions. 2-Methoxyethanol was dried over a molecular sieve and distilled prior to use. The desired amounts of starting materials with 5mol% excess Bi and Pb were dissolved in absolute 2-methoxyethanol, and the addition of excess Bi and Pb was performed to compensate for the loss due to the Bi and Pb volatilization during the heat treatment of precursor powders and films. Thin films were fabricated using the precursor solution by spin-coating on $Pt/TiOx/SiO_2/Si$ substrates.

To study the crystallization and thermal behavior of BS-PT powders, the precursor powders of BS-PT were prepared by removing the solvent from the precursor solutions. These powders were heat treated at 800°C in an oxygen atmosphere to yield the crystalline BS-PT powders. The crystallographic phases of the BS-PT powders and thin films were characterized by X-ray diffraction analysis (XRD). The thermal behavior of the crystallized BSPT powder was analyzed by differential scanning calorimetry (DSC) in air from room temperature to 600°C with a temperature ramp rate of 20°C/min. The surface morphology of the crystallized thin films was observed using atomic force microscopy (AFM). The electrical properties of the films were measured using a 0.2-mm diameter Pt top electrode deposited by DC sputtering onto the surface of the BS-PT films followed by annealing at 400°C for 1 hour. The Pt layer of the Pt/TiOx/SiO2/Si substrate was used as the bottom electrode. The electrical properties of the films were evaluated using an impedance/gain-phase analyzer and a ferroelectric test system at room temperature.

Crystallization and Phase Transition of BS-PT Powders

By reacting starting metal-organic compounds in 2-methoxyethanol, homogeneous $BiScO_3\text{-}PbTiO_3$ precursor was successfully prepared. In the BS-PT system, the MPB region has been confirmed to be around x = 0.64 in the $(1\text{- }x)BiScO_3\text{-}xPbTiO_3$ system, by analyzes of the detailed X-ray and electron-diffraction patterns, dielectric properties and DSC data of BS-PT single crystals and polycrystals. The excess Bi and Pb concentration of the precursor powders was determined when the perovskite BS-PT phase having the highest crystallinity without the formation of second phase was obtained. At compositions of x = 0.5 and x = 0.6, the BS-50PT and BS-60PT precursor powders crystallized in the rhombohedral (pseudo cubic) perovskite phase, which is consistent with the BS-PT with x = 0.60 that had a rhombohedral perovskite BS-PT phase. At x = 0.7 (BS-70PT) composition, the main crystallographic phase was considered to be the tetragonal BS-PT phase because of the separation of 002 and 200 diffractions in the XRD pattern. Furthermore, at x = 0.65, the BS-65PT precursor powder might crystallize in a mixture of rhombohedral and tetragonal BS-PT phases. The MPB of the chemically derived BS-PT powders was considered to be located at around the x = 0.65 composition.

A DSC curve was illustrated for the BS-65PT powder from room temperature to 600°C. To confirm the ferroelectric phase transition behavior of BS-65PT (at MPB composition), these DSC measurements were conducted, since DSC or differential thermal analysis (DTA) is useful for detecting the exo- and endothermic behavior of ferroelectrics during phase transition. The phase transition from ferroelectrics to paraelectrics was endothermic. The Curie temperature of BS-65PT was determined to be 470°C, which is almost the same as the temperature of bulk BS-PT ceramics, although the DSC curve was broadened. Thus, chemically synthesized BS-65PT was ferroelectric over a wide temperature range, and maintained the high Curie temperature of bulk BS-PT ceramics.

The BS-65PT thin film had the highest dielectric permittivity of approximately 1700 at room temperature. The BS-65PT thin film also showed the highest values of ferroelectric properties with a Pr of 23 ¯C/cm2 and an Ec of 70 kV/cm at an applied field of 400 kV/cm. The existence of the MPB composition at around x = 0.65 contributed to these results.

Ferroelectricity in Nanostructured $BiScO_3$–$PbTiO_3$ [52]

By spark plasma sintering of nanocrystalline powders obtained by mechanosynthesis, nanostructured ceramics of high-temperature piezoelectric $0.375BiScO_3$–$0.625PbTiO_3$ were prepared. The electrical properties of dense ceramics with decreasing average grain size down to 28 nm were characterized. The electric field is screened by the electrically insulating grain boundaries at the nanoscale, in which size effects in ferroelectric polycrystalline materials need to be considered. Moreover, the requirement of increasingly large electric fields to achieve a given polarization with the decrease in the grain size, together with the depletion of the dielectric anomaly associated with the ferroelectric transition until its disappearance, seems to be the contribution of the grain boundary effects.

The fabrication of piezoelectric devices that meet the current miniaturization trend in the ceramic technologies for microelectronics, demands functional materials with submicrometre or even nanometric grain size for reliability. The feasibility of the miniaturized technologies depends on how functional properties, such as polarization switching, retention and piezoelectricity, behave with the grain size reduction to the nanoscale. This raised the interest in studying grain size effects in ferroelectric polycrystals, such as perovskite morphotropic phase boundary (MPB) materials, as they have shown very high piezoelectric activity. These materials are good candidates to implement piezoelectric sensing and actuation in a range of anticipated micro- and nanotechnologies.

Previous studies in MPB materials were focused on relaxor-based solid solutions with the highest piezoelectric coefficients reported in single crystals. The stabilization of the high-temperature relaxor state for nanostructured ceramics has been found, indicating that the long-range ferroelectric order does not develop in the nanoscale. Thus, a similar in-depth study of a non-relaxor MPB material with high piezoelectric activity down to the nanoscale needs to be studied.

As compared with $Pb(Zr, Ti)O_3$, $BiScO_3$–$PbTiO_3$ is a MPB material that exhibits high Curie temperature (T_C) and large piezoelectric coefficients (d_{33}). The broad range of used temperatures (~350 ◦C) makes this material promising for the next generation of high temperature and high-power piezoelectric devices. The MPB composition region of $(1-x)BiScO_3$–$xPbTiO_3$ solid solution is as wide as $0.6 < x < 0.7$, in which monoclinic *Cm* and tetragonal *P4mm* polymorphs coexist. Furthermore, a recent work indicates that *Cm* symmetry persists in very small size nanocrystals, and large piezoelectric properties have been addressed in fine-grained ceramics in which grain growth was restricted to the submicrometre range.

MPB $BiScO_3$–$PbTiO_3$ nanostructured ceramics shown polarization switching and radial piezoelectric resonance after poling, thus demonstrating macroscopic functionality and better down-scaling behavior than that of relaxor-based MPB materials. Macroscopic properties of $0.375BiScO_3$–$0.625PbTiO_3$ (BSPT) ceramics with decreasing grain size from the submicrometre range down to the nanoscale, which clearly indicate that grain size effects alone cannot account for the observed phenomenology, and that grain boundaries may play a dominant role for the smallest sizes.

Highly dense, submicrometre- and nanostructured MPB region ($x = 0.625$) BSPT ceramics were prepared by spark plasma sintering, and the nanocrystalline powders were obtained by mechanosynthesis. These powders were obtained by mechanochemical activation of the binary oxides in oxygen atmosphere using a high-energy planetary mill. The transmission electron microscopy (TEM) measurements show an average particle size of 11 nm for these powders. SPS conditions were tailored for obtaining ceramics with a range of grain sizes in the submicrometre and nanometric ranges. This technique allows high densification at moderate temperatures and short times, but limit grain growth. Results shown here correspond to ceramics sintered at 650 and 800 ◦C under uniaxial pressures of 75 and 120MPa during 3 minutes. In addition, results on a coarse grain ceramic (CGC) prepared from the same powder by conventional sintering at 1100 ◦C for 1 hour are also shown for comparison.

A typical TEM image shows the grain size distribution of the BSPT ceramic with an average grain size of 28 nm. To obtain reliable size distributions, Feret's diameters of more than 150 grains were measured on TEM or scanning electron microscopy images of the nanostructured and submicrometre-grain ceramics.

From the temperature dependence of the relative dielectric permittivity *(K)* and losses (tan δ) for the SPS ceramics under study at several frequencies, a low-frequency dispersion was observed in the temperature range where the dielectric anomaly associated with the transition from the ferroelectric to the paraelectric state occurs for this composition. $TC \sim 465$ ◦C; which is taken into account that the high-temperature dielectric response of ferroelectric polycrystals usually consists of several contributions that include Debye-type relaxation processes, often appearing in perovskite-type oxides and are attributed to defects in the materials, e.g. short range hopping of oxygen vacancies. Reducing conditions within the graphite die during the SPS experiment might promote the formation of these defects. Nevertheless, a size effect in the ferroelectric transition is evident in the curves obtained at 1 MHz, a frequency at which the relaxation has shifted well above T_C. The dielectric anomaly associated with the ferroelectric transition was 442 ◦C for the nanostructured ceramic which was strongly flattened and broadened.

The ferroelectric transition cannot be discerned from the curve of the nanostructured ceramic with the smallest grain size. The *P–E* hysteresis loops for the ceramics under study along with that obtained for the CGC for comparison. Asquare loop close to saturation was obtained for the CGC, as confirmed by the curves of spontaneous (*Ps)* and remanent polarization (*Pr)* against the maximum applied electric field.

Macroscopic ferroelectricity was also found in the nanostructured ceramics with decreasing grain size down to 80 nm. The polarization switching was not observed for the material with 28 nm, but a typical linear dielectric behavior can be observed. Both spontaneous and remanent polarizations continuously decrease with the decrease in the grain size and the loops become increasingly lean. The saturation was far from the case of the nanostructured ceramics. The electric field derivative of the spontaneous polarization continuously decreases above 40 kV cm−1 for the CGC, indicating that saturation is slowly approached, but it is still increasing at the highest measured electric fields for all the SPS ceramics. These derivatives strongly suggest that materials progressively move away from saturation with the decrease in the grain size.

The occurrence of macroscopic ferroelectric switching indicates that the nanostructured ceramics can be poled, and their piezoelectric properties can be studied. Significant variations of the d_{33} coefficient were found with decreasing grain size. On the one hand, a relatively high d_{33} of 110 pCN^{-1} was obtained for the ceramic with size of 375 nm. However, a low d_{33} of 24 pCN−1 was measured for the ceramic with size of 80 nm, and only 2 pCN−1 was achieved for the ceramic with size of 28 nm. These values show the reduced but measurable piezoelectricity of the nanostructured ceramics.

The grain size effects on the spontaneous polarization and the d_{33} coefficient are accompanied by the fact that the total resistivity (ρ) of the SPS ceramics significantly increases with the decrease of the grain size. Since in electroceramics and typically in ferroelectrics, the resistance of grain boundaries exceeds that of the grains, this case expected an increase in the total resistivity for nanostructured ceramics, due to their larger volume fraction of insulating grain boundaries. This seems to be the case for the BSPT ceramics under study in this part. Moreover, very large resistivity was measured for the material which is two orders of magnitude larger than that obtained for the CGC, suggesting an even larger contribution of the electrically insulating grain boundaries to the total resistivity for this ceramic.

The temperature dependence of the total dc conductivity (dc*)* for the ceramics under study were obtained by circuit analysis of the impedance spectroscopy data. Similar Arrhenius dependences were observed for the CGC and the nanostructured ceramics, for which a change in slope at ~300 ◦C is observed. Above this temperature, the activation energy (*Ea)* was ~1.1 eV, typical of mobile doubly ionized oxygen vacancies ($V^{\cdot\cdot}O$), which can also be responsible for the dipolar high-temperature relaxation observed in the dielectric curves. Below 300 ◦C, a change of activation energy in the Arrhenius behavior was ~0.6 eV.

The ac conductivity (ac*)* of all ceramics displays two contributions at low and high frequencies in the temperature range between 200 and 500 ◦C, which are usually associated with the grain boundary and bulk responses, respectively, and indicating that the grain boundaries (low-frequency contribution) dominate the total dc conductivity.

The temperature dependence of the total dc conductivity for the nanostructured ceramic with the smallest grain size shows a different behavior. The activation energy is ~1.4 eV in all the temperature ranges. The higher activation energy and the lower total dc conductivity obtained for this ceramic may be attributed to a mechanism of association of defects at the grain boundaries, which reduces the mobility of the doubly ionized oxygen vacancies, indicating that a more resistive grain boundary is dominating the total response, and is consistent with the exceptionally high resistivity calculated from the hysteresis loops.

When decreasing grain size, to achieve saturation of polarization, increasingly large electric fields are required. This might be attributed to a grain boundary effect, and so might be the depletion of the dielectric anomaly associated with the ferroelectric transition. The grain boundary effect must be also considered to understand the mechanisms behind the behavior of the dielectric permittivity for the nanostructured ceramic with the smallest grain size (28 nm). The dielectric anomaly vanishes, either because of the disappearance of the ferroelectric transition or because of its severe depletion, considering the exceptionally high grain boundary resistivity of this ceramic.

Twin-free $BiScO_3$-$PbTiO_3$ Nanocrystals [53]

MPB ferroelectric perovskite solid solutions have been studied extensively as the best candidates for electromechanical transduction due to their exceptionally high piezoelectric coefficients. However, there is still ambiguity about the actual symmetry of the phases at the MPBs. A detailed analysis of the crystalline structure of $0.39BiScO_3$-$0.61PbTiO_3$ twin free nanocrystals, using the combination of aberration-corrected high resolution electron microscopy and multislice simulations, indicated the presence of a monoclinic symmetry for crystals with an average size of 15 nm, which is strong evidence in favor of the actual existence of the polymorph at the MPB. It also shows that in nanocrystals, perovskite distortion is still present, which is important from the point of view of the size limit of ferroelectricity and consistent with the electrical measurements in nanostructured ceramics.

Ferroelectric perovskite solid solutions with a morphotropic phase boundary MPB between ferroelectric polymorphs of rhombohedral and tetragonal symmetries present the highest known piezoelectric coefficients. Recently, several works reported the existence of monoclinic phases with space groups *Pm* and *Cm* at the MPBs of the PZN-PT and PMN-PT systems. However, the monoclinic phases existence was questioned after the experimental observation of a significant decrease in the ferroelectric/ferroelastic domain size in PZT at the MPB down to the nanoscale , and of a correlation between the nanoscale domain configurations and the monoclinic distinctive features in the diffraction data. It followed a demonstration by diffraction theory that nanotwin superlattices of either rhombohedral or tetragonal phases give place to effective adaptive monoclinic phases of space groups *Cm* and *Pm*, respectively, with cell parameters that are intrinsically related to those of the rhombohedral and tetragonal phases. In this model, an effective polarization rotation can still take place under the electric field by rearrangement of the nanodomains.

Nevertheless, these latter experimental and theoretical results do not imply that the monoclinic phases do not exist. The controversy can be solved by studying the actual symmetry of a twin free crystal at the MPB. Indeed, local techniques derived from the transmission electron microscopy are suitable to determine without ambiguity the symmetry of small nanodomains. By using convergent beam electron-diffraction CBED, $PbZr_{0.54}Ti_{0.46}O_3$ ceramics with domains of 30–100 nm width were investigated, which indicated that the average crystal symmetry of the probed volume had to be monoclinic. However, it is well known that CBED techniques cannot be used for symmetry determination in very thin crystal, where dynamical effects are not sufficient to provide clear symmetry in the diffraction disks.

By using aberration-corrected high resolution electron microscopy (ACHREM), the crystal symmetry of twin free nanocrystals of MPB $0.39BiScO_3$-$0.61PbTiO_3$ composition as small a size as 10–30 nm in diameter was studied. The nanocrystals were chosen, since it is well known that twin formation is energetically unfavorable below a critical size. ACHREM is an advanced microscopy technique that has allowed a step forward in qualitative and quantitative electron microscopy, providing genuine atomic resolution, so there is no doubt that it opens a range of novel possibilities in materials.

Perovskite phase $0.39BiScO_3$-$0.61PbTiO_3$ nanocrystals were obtained by the mechanosynthesis method. The mechanochemical treatment was carried out in oxygen in order to avoid the Bi_2O_3 reduction, for times up to 35 hours. Crystallographic evolution during mechanical activation of the initial mixture was investigated by x-ray powder diffraction XRD. From the XRD pattern of the final product of the mechanochemical activation process from the stoichiometric mixture, the $0.39BiScO_3$-$0.61PbTiO_3$ perovskite was mechanosynthesized and isolated as a single phase after the mechanical treatment. Mechanosynthesis is a suitable preparation route to obtain nanocrystalline oxides with perovskite-type structure. The crystal size calculated from the XRD data using the Scherrer formula is about 10 nm, which was confirmed by conventional TEM, providing reliable size distributions for such nanocrystalline powders. The method for the preparation of the TEM studies the powder sample: first it was crushed in an agate mortar and suspended in *n*-butanol. After ultrasonic dispersion, a droplet was deposited on a copper grid supporting a perforated carbon film.

In order to carry out the advanced structural characterization, mechanosynthesized perovskite was also investigated by ACHREM. The microscope conditions were set and measured by the corrector software using the classical Zemlin-tableau method. All of the experiments were performed at 100 kV due to the high instability of the sample using a high-energy electron beam.

Conventional high-resolution transmission electron microscopy is a very efficient method for studying the structural properties of crystals at the nanoscale. However, a fine interpretation of the HREM contrast in terms of symmetry, atomic positions, etc., of the crystal requires control of the microscope's imaging conditions. Indeed, the atomic fringes contrast can be strongly affected by the phase modification of the diffracted beam coming from the uncorrected objective. To overcome these difficulties, focal series of HREM pictures are normally implemented in order to calculate and suppress numerically these phase perturbations, using dedicated software. In the case of ACHREM, we can have a perfect control of the high-order aberrations coefficients which furthermore allowed us to map the phase modification added by the system with the residual aberrations. Focal series should be implemented to suppress aberrations up to fifth-order astigmatism. Concerning our sample, we have found that these high-order aberrations have not a sufficient effect to modify the atomic contrast observed.

The use of aberration correction for imaging nanocrystalline samples has some important advantages: 1.) it minimized delocalization of the image contrast, which results in the observation of fine details coming from the high spacial frequencies components of the diffraction pattern and 2.) enhanced image contrast and field of view by correcting off-axial coma. As it is known, comparison between experimental fringes contrast and simulated ones for a defined structure is very problematic, using conventional HREM due to the large amount of parameters which must be involved in the simulation.

The corrector could be easily included in the simulation. In order to retrieve the crystalline structure of the different twin free nanocrystals, multislice simulations of several ACHREM pictures of the mechanosynthesized perovskite were performed using different input structures based on rhombohedral *R*3*m*, tetragonal *P*4*mm* and monoclinic *Cm*.

An example of the Zemlin tableau can be used to determine the value of high-order aberrations inserted in the selected multislice simulations. The thickness has been determined assuming a spherical shape of each nanocrystal with a lateral size of 10–30 nm. Images with a remarkable high contrast and resolution were obtained, leading to high quality fast Fourier transform FFT for each individual nanocrystal, even when it was necessary to set the corrector and the microscope at 100 kV acceleration voltage to avoid degradation phenomena under the beam.

Through the comparison between the high quality fast Fourier transform FFT and the simulated electron-diffraction pattern of the three possible polymorphs tetragonal, rhombohedral, and monoclinic it was possible to retrieve the crystallographic orientation zone axis of the structures under analysis. The HREM contrast taken on a particle of about 20 nm in size, shows the FFT belonging to the nanosized crystal from which the zone axes of each polymorph were obtained: (001) in the case of the tetragonal *P*4*mm*, (111) for the rhombohedral *R*3*m* and (011) for the monoclinic *Cm*. The main advantage of using ACHREM comes from the possibility to observe the fine structure of the contrast, coming from the contribution of high spatial frequencies of the diffraction pattern. Indeed, in the aberration-free conditions, the transfer function of the microscope stays flat even for high spatial frequencies components without the contrast inversion usually observed in an uncorrected microscope. All of the spatial frequencies are equally transmitted by the optical system without any delocalization. As a result, due to the contribution of these diffracted beams, very small interference fringes, directly related to the object fine atomic structure, induced fine details, can be observed. The secondary spots can be directly compared with the simulated contrast performed by the multislice algorithm using the chosen structure and the microscope conditions given by the corrector. In the case of the nanocrystal, it is possible that only in the monoclinic symmetry, the fine structure of the simulation fits perfectly.

These clearly revealed the monoclinic symmetry of twin free nanocrystals in the $BiScO_3$-$PbTiO_3$ solid solution, and they are strong evidence that monoclinic phases at MPBs are not the result of the nanotwin superlattices of the other polymorphs. The results reported here are also relevant to another highly topical issue: the size limit for ferroelectricity. This is a cooperative phenomenon, and thus the existence of a fundamental size limit below which ferroelectricity vanishes is expected. In relation to MPB perovskites, the predominant tetragonal phase has been reported for PZT nanoparticles of 30 nm by Rietveld analysis of XRD data. Unlike in PZT, ferroelectric order does not develop in nanoscale relaxor based materials, and instead the high-temperature relaxor state is stabilized. In the case of MPB $BiScO_3$-$PbTiO_3$, monoclinic symmetry is found for nanocrystals as small as 10 nm.

These results compare favorably with those reported for relaxor based MPB systems, for which the high-temperature relaxor state was stabilized at room temperature as a side effect, and no ferroelectric long-range order develops.

REFERENCES

[1] D. B. Chrisey and G. K. Hubler, *Pulsed Laser Deposition of thin films*, (1994), ISBN 0-471-59218-8.

[2] R. W. Jones, *Fundamental principles of sol-gel technology*, (1989) ISBN 0-901462-69-1.

[3] N. Braithwaite, G. Weaver, *Electronic materials,* (1990) ISBN 0-408-02840-8

[4] Hyundai Electronic Industries, URLhttp://kcs.hei.co.kr/models/fram/fram.html.

[5] S. B. Desu, P. C. Joshi, X. Zhang and S. O. Ryu, *Appl. Phys. Lett*., 71(1997), S1041.

[6] A. I. Kingon, S. K. Streiffer and S. H. Kim, *Dram's and Fram's based on ferroelectric films*, Tutorial Paperback ISAF'98.

[7] F. Wang and S. Leppävuori, *J. Appl. Phys*., 82(1997), S1293.

[8] M. Grossman et al., *Appl. Phys. Lett*., 76(2000), S363.

[9] S. Aggarwal et al., *J. Appl. Phys*., 83(1993), S1617.

[10] S. Sadashivan, S. Aggarwal, T. K. Song and R. Ramesh, *J. Appl. Phys*., 83(1998), S2165.

[11] N. Poonawala, V. P. Dravid, O. Aiciello, J. Im and A. R. Krauss, *J. Appl. Phys*., 87(2000), S2227.

[12] X. Dingquan and J. Kor. *Phys. Soc*., 32 (1998), S1798.

[13] A. Pignolet, C. Curran, M. Alexe, S. Senz, D. Hesse, U. Gösele, *Integrated Ferroelectrics,* 21(1998), S485.

[14] Z. Huang, Q. Zhang and R. W. Whatmore, *J. Appl. Phys*., 86(1999), S1622

[15] D. Peroulis, S.P. Pacheco, K. Sarabandi, and L.P.B. Katehi, *IEEE Trans. Microwav. Theory Tech.*, 51(10), 259 (2003).

[16] F. Xu, S. Trolier-McKinstry, W. Ren, and B. Xu, *J. Appl. Phys.*, 89(2), 1336 (2001).

[17] K. Saito, T. Kurosawa, T. Akai, T. Oikawa, and H. Funakubo, *J. Appl. Phys.*, 93(1), 545 (2003).

[18] X.H. Du, J.H. Zheng, U. Belegundu, and K. Uchino, *Appl. Phys. Lett.*, 72(19), 2421 (1998).

[19] S.-E. Park and T.R. Shrout, *IEEE Trans. UFFC*, 44, 1140 (1997).

[20] B. Noheda, D.E. Cox, G. Shirane, R. Guo, B. Jones, and L.E. Cross, *Phys. Rev. B*, 63(1), 014103 (2001).

[21] H.D. Chen, K.R. Udayakumar, C.J. Gaskey, and L.E. Cross, *Appl. Phys. Lett.*, 67(23), 3411 (1995).

[22] A. Seifert, N. Ledermann, S. Hiboux, J. Baborowski, P. Muralt, and N. Setter, *Integr. Ferro*., 35(1–4), 1889 (2001).

[23] F. Xu, R.A.Wolf, T. Yoshimura, and S. Trolier-McKinstry, *Proc. 11th Int. Symp. Electrets*, 386 (2002).

[24] R.A.Wolf and S. Trolier McKinstry, *J. Appl. Phys.*, 95(3), 1397 (2004).

[25] T. Haccart, C. Soyer, E. Cattan, and D. Remiens, *Ferroelectrics*, 254(1–4), 185 (2001).

[26] I. Kanno, H. Kotera, K.Wasa, T. Matsunaga, T. Kamada, and R. Takayama, *J. Appl. Phys.*, 93(7), 4091 (2003).

[27] D.-J. Kim, J.-P. Maria, A.I. Kingon, and S.K. Streiffer, *J. Appl. Phys.*, 93, 5568 (2003).

[28] R.E. Eitel, C.A. Randall, T.R. Shrout, P.W. Rehrig, W. Hackenberger, and S.E. Park, *Jpn. J. Appl. Phys. Part 1*, 40(10), 5999 (2001).

[29] T.R. Shrout, Private Communication (2002).

[30] K. Kakimoto, H. Kakemoto, S. Fujita, and Y. Masuda, *J. Am. Ceram. Soc.*, 85(4), 1019 (2002).

[31] B.A. Tuttle, T.J. Garino, J.A.Voight, T.J. Headley, D. Dimos, and M.O. Eatough, *Science and Technology of Electroceramic Thin Films*, edited by O. Auciello and R. Waser, Kluwer Academic Publishers, The Netherlands, 117 (1995).

[32] SusanTrolier–Mckinstry, et al, J. Appl. Phys., Vol. 92, No. 7, 1 October 2002.

[33] R. E. Eitel, C. A. Randall, T. R. Shrout, P. W. Rehrig, W. Hackenberger, and S. E. Park, *Jpn. J. Appl. Phys*., Part 1 40, 5999(2001).

[34] Takeshi Yoshimura and Susan Trolier-McKinstry. *Appl. Phys. Lett*., Vol. 81, No. 11, 9 September 2002.

[35] Hai Wen, Xiaohui Wang, and Longtu Li. *J. Am. Ceram. Soc*., 89 [7] 2345–2347 (2006).

[36] R. E. Eitel, C. A. Randall, T. R. Shrout, and S. E. Park, ''Preparation and Characterization of High Temperature Perovskite Ferroelectrics in the Solid- Solution (1-x)BiScO3–xPbTiO3, " *Jpn. J. Appl. Phys*., 41 [4A] 2099–2104 (2002).

[37] Hai Wen, Xiaohui Wang, Xiangyun Deng, and Longtu Li. *Appl. Phys. Lett*. 88, 222904 (2006).

[38] Hai Wen, Xiaohui Wang, and Longtu Li. *J. Appl. Phys*. 100, 084315 (2006).

[39] *J. Inorg. Nucl. Chem*. 9, 113 (1959).

[40] Hai Wen, Xiaohui Wang, Caifu Zhong, Like Shu, and Longtu Li. *Appl. Phys. Lett*. 90, 202902(2007).

[41] C. F. Zhong, X. H. Wang, H. Wen, L. T. Li, C. W. Nan, and Y. H. Lin. *Appl. Phys. Lett*. 92, 222910 (2008).

[42] Takeshi Yoshimura and Susan Trolier-McKinstry. *Appl. Phys. Lett.,* Vol. 81, No. 11, 9 September 2002.

[43] Hai Wen, Xiaohui Wang, and Longtu Li. *J. Am. Ceram. Soc*., 89 [7] 2345–2347 (2006).

[44] Hai Wen, Xiaohui Wang, Xiangyun Deng, and Longtu Li. *Appl. Phys. Lett*. 88, 222904 (2006).

[45] Hai Wen, Xiaohui Wang, Caifu Zhong, Like Shu, and Longtu Li. Appl. Phys. Lett. 90, 202902(2007) C. F. Zhong, X. H. Wang, H. Wen, L. T. Li, C. W. Nan, and Y. H. Lin. *Appl. Phys. Lett*. 92, 222910 (2008).

[46] Hai Wen, Xiaohui Wang, Caifu Zhong, Like Shu, and Longtu Li. *J. Am. Ceram. Soc.,* 90 [8] 2441–2445 (2007).

[47] S. E. Park and T. R. Shrout, *J. Appl. Phys*. 82, 1804 (1997).

[48] Hai Wen, Xiaohui Wang, and Longtu Li. *J. Am. Ceram. Soc*., 90 [10] 3248–3254 (2007).

[49] Hai Wen, Xiaohui Wang, Xiangyun Deng, Tieyu Sun, and Longtu Li. *Journal of Applied Physics* 101, 016103 (2007).

[50] Bog G. Kim et al, *Journal of Applied Physics* 109, 064109 (2011).

[51] Xiaohui Wang et al, *Ceramics International* 38S (2012) S237–S240.

[52] SusanTrolier–Mckinstry et al, *Applied Physics Letters* 95, 162901 2009.

[53] Hiroshi Yamazaki et al, *Journal of the Ceramic Society of Japan* 118 [8] 631-635 2010.

[54] H Amorın et al, *J. Phys. D: Appl. Phys*. 43 (2010) 285401.

[55] Teresa Hungría et al, *Physical Review* B 81, 100102R 2010.

Chapter 6

TEXTURE ENGINEERED BSPT THIN FILMS

Temperature stability of the electric properties, as well as good ferroelectric or piezoelectric activity, is important for MEMS device application. Ferroelectric thin films or ceramics with high Curie temperatures and high orientation are very attractive in many fields. This research is aimed at the development of a reliable processing method for preparation of alternative relaxor ferroelectrics with high Curie temperatures and high orientation (textured), and focused on the chemical solution fabrication of the $BiScO_3$-$PbTiO_3$ (BS-PT) thin film system.

In this work, $BiScO_3$-$PbTiO_3$ (BS-PT) thin film system was selected as a representative material, instead of $Pb(In_{1/2}Nb_{1/2})O_3$-$PbTiO_3$(PINT) and $Pb(Yb_{1/2}Nb_{1/2})O_3$-$PbTiO_3$ (PYNT) systems, because it holds excellent electric properties and a much higher Curie temperature. The recent development of high Curie temperature BS-PT ferroelectrics were simply reviewed and some results of preparing and characterizing BS-PT ferroelectric thin films with highly texture by chemical solution deposition (CSD) approach were introduced.

Several stable BS-PT precursor solutions without aqueous solution and addition of toxic methoxyethanol have been successfully prepared, which solves a key issue in fabricating thin films and makes the CSD deposition of BS-PT films feasible. Applying the spin coating method, we successfully prepared BS-PT thin films on the substrates of $Pt/Ti/SiO_2/Si$ with different composition, and the effect of the spin rate and spin time on the film quality has been investigated. With appropriate post-coating thermal processes, pure perovskite BS-PT phases have been obtained. It is indicated that these films show (100)-oriented texture, and furthermore, it is discovered that the crystalline temperature plays an important role in controlling the texture degree. In addition, by introducing a seed layer of $PbTiO_3$, PbO, or TiO_2, the (100)-oriented texture degree will be greatly enhanced. Other factors, such as film thickness and solutions concentrations, were also found to influence the texture controlling. The surface and cross-section morphology of BS-PT thin films has been obtained by the SEM techniques, from which the thin film's grain size and thickness is estimated.

The main problem in this work is that the quality of BS-PT thin films is not good enough to obtain ultra-high dielectric constant. For future studies, we plan to overcome this shortage through the approaches of further improving wettability between the precursor solution and substrates, further optimizing the thin film growth parameters, increasing the orientation degree of BS-PT thin films, etc.

It is evident that the CSD method is a reliable processing approach for fabricating the high texture BS-PT films, which is expected to improve the piezoelectric properties with high-Tc. The results obtained seem to be novel and of scientific significance, which are worthy for further investigation and exploration in the next research stage.

6.1. Development of MPB $BiScO_3$-$PbTiO_3$ Films

1. Background and Motivation

Oxide materials with the ABO_3 perovskite structure, such as Pb(Zr1-xTix)O_3 (x~0.48) (PZT) system, have become widespread in industry applications [1-5]. Recent developments in regard to relaxor-based single crystal piezoelectrics, such as $Pb(Zn_{1/3}Nb_{2/3})O_3$– $PbTiO_3$ (PZNT) and $Pb(Mg_{1/3}Nb_{2/3})O_3$– $PbTiO_3$ (PMNT) have shown superior performance characteristics, such as their ultrahigh piezoelectric properties, compared to Pb(Zr, Ti)O_3 (PZT) polycrystalline ceramics [6-9]. The enhanced piezoelectric activity of PZNT and PMNT, however, comes with the sacrifice of the temperature stability of the properties, being limited to their relatively low Curie temperatures (T_C≈178 °C), and further restricted by their lower ferroelectric phase transition temperatures (*Tr-t*≈ 120 °C). In some important applications, especially in the automotive and aerospace industries, the need for actuation and sensing over a broad temperature range is essential. Therefore, the low Curie temperature and relatively low ferroelectric phase transition temperatures (rhombohedral to tetragonal phase) of PZNT and PMNT systems limit their implementation in many fields. On the other hand, the temperature stability of the electrical properties, as well as good ferroelectric or piezoelectric activity, is important for device applications. So, ferroelectrics with high transition temperatures (*Tc)* are attractive [10-13].

Many relaxor-PbTiO3 ferroelectrics, such as $Pb(In_{1/2}Nb_{1/2})O_3$-$PbTiO_3$(PINT), $Pb(Yb_{1/2}Nb_{1/2})O_3$-$PbTiO_3$ (PYNT), $BiScO_3$-$PbTiO_3$ (BSPT) systems, have Curie temperatures as high as that of PZTs's or about 100°C higher than that of PZTs, and hold excellent properties, which can overcome the shortage of PZNT and PMNT crystals, and meet the requirements in applications as next generation materials for actuators in automotive and aerospace sensors.

The BS-PT system is based on the general solid solution formula (1-x)$BiMeO_3$-x$PbTiO_3$, where Me is a suitably sized 3^+cation (Sc, In, Yb, etc.). Based on the tolerance trends in perovskite ferroelectrics, a relationship between tolerance factor (t) and Curie temperature(Tc) was proposed, namely, the tolerance factor can be used to estimate the changes of Curie temperature with different perovskite systems [1, 14-17]. From the empirical analysis, $BiMeO_3$-based perovskites with decreasing values (t) were suggested to confirm the premise that Tc increases with a decreasing tolerance factor. It is evident that a solid solution of $BiScO_3$(t=0.907), $BiInO_3$(t=0.884), or $BiYbO_3$(t=0.857) with $PbTiO_3$ would exhibit MPBs with Tc were significantly greater than PZT(t=0.96, Tc=386°C) [29]. Solid solutions with $BiMe^{+3}O_3$-$PbTiO_3$ have small tolerance factors and are proving to give new high temperature ferroelectric solid solutions. Although, the Tc of $BiInO_3$-$PbTiO_3$ and $BiYbO_3$-$PbTiO_3$ systems may reach 560°C and 700°C, respectively, the instability of the peroveskite phase in these two materials prevent them from being easily synthesized. For (1-

x)$BiInO_3$-x$PbTiO_3$, pure perovskite phases could not be achieved even for x≧0.9. To enhance the perovskite stability, the addition of 1-3 mole% $BaTiO_3$ was introduced into the system for the compositions with x =0.7 [1]. Though the existence of a MPB was not confirmed, the dielectric temperature behavior for a tetragonal composition (x=0.7) revealed a Tc = 570°C, being significantly higher than that of PZT and $PbTiO_3$ themselves. The transition temperature-tolerance factor relationship was further substantiated in the (1-x)$BiYbO_3$-x$PbTiO_3$ system, though the sample was a mixed phase. [1]

In this work, we focus on the (1-x)BiScO3-x$PbTiO_3$ system, exhibiting a MPBs and having a Tc greater than 450°C. Recently, the (1-x)$BiScO_3$-x$PbTiO_3$ solid solution system has demonstrated the potential for producing piezoelectrics with higher piezoelectric activity, higher in-use temperatures, and lower temperature-property dependence than any present-day ferroelectric material, with the perovskite or any other crystal structure [18-21].

$BiScO_3$ (BS) ceramics were first synthesized in 1969 [22], under a high pressure by Tomashpol'skii et al. Under ordinary pressure, pure $BiScO_3$ is unstable. But $(1-x)BiScO_3$-$xPbTiO_3$ (BS-PT) exhibits a temperature stability of the electrical properties until 350 °C and has a T_C about 450 °C. The phase diagram of the BS-PT system indicates that, at room temperature, the rhombohedral symmetry lies in the region of 50<x<62mol% $PbTiO_3$ (PT), the tetragonal phase lies in the range x>65mol% PT. In the solid-solution range 62<x<65mol% PT, there exists a MPB phase [1, 23].

BS-PT ceramics exhibit good piezoelectric properties and a high ferroelectric transition temperature comparable to PZT ceramics near the MPB. Using conventional solid state processing techniques, rhombohedral, MPB, and tetragonal phase BSPT ceramics (at approximately 62-65%$PbTiO_3$) have been prepared respectively [1, 21, 24-25]. In the composition of x=0.64, especially, high piezoelectric properties, d_{33}=460pC/N and high curie temperature, Tc=450°C is obtained. Results of high temperature piezoelectric measurements show that high piezoelectric coefficients persist to temperatures approaching 400°C.

In order to improve the piezoelectric properties of BS-PT system, Zhang et al [10, 26-29] performed the investigations on growth and characterization of BS-PT single crystals. They focused their attention on the composition of 0.43$BiScO_3$– 0.57$PbTiO_3$. BiScO3-PbTiO3 (BSPT) single crystals have been grown by the flux method. The piezoelectric properties of (001) 0.43BiScO3– 0.57PbTiO3 single crystals can be found within the papers of Zhang SJ, which can reach the order of those of relaxor-$PbTiO_3$ single crystals, but their usage temperature range for their application are broader. The enhanced properties in the BS-PT crystal make this material a promising candidate for high temperature and temperature stable actuators and transducers. However, it is difficult to grow large size and good quality single crystals, and the cost of crystal growth is considerably higher than that of growing ceramics and thin films.

It is important to notice that the same properties also remain preserved in BS-PT thin films [24]. Since temperature stability of the electric properties, as well as good ferroelectric or piezoelectric activity, is important for MEMS device application, ferroelectric thin films with high transition temperatures are attractive. Micro-Electro-Mechanical Systems (MEMS) is the integration of mechanical elements, sensors, actuators, and electronics on a common silicon substrate through microfabrication technology. Electromechanical actuators directly transform input electrical energy into mechanical energy. T. Yoshimura et al [30] reported that 1-μm thick epitaxial films of (001) $BiScO_3$-$PbTiO_3$ near MPB had been grown on (100) $SrRuO_3$/(100) $LaAlO_3$ substrates by pulsed laser deposition from a $BiScO_3$-$PbTiO_3$ (40/60 or

50/50) ceramic target. The films had a room temperature dielectric constant of 850-800, tanδ = 0.08-0.09, and maximum dielectric constant of 5530-3600 at 455-460 °C. The remanent polarization is about 42-36 μC/cm^2 and the coercive field was 75 kV/cm. The effective transverse piezoelectric coefficient e_{31f} was –9 to -12 C/m^2. This result is quite encouraging for sensor and actuator device development because the observed piezoelectric properties are as good as (001) oriented Pb(Zr, T)O_3 films ($e_{31,\ f}$ ~ -12 C/m^2), while the transition temperature is 100 °C higher than that of PZTs.

This research was aimed at the development of a reliable processing method for preparation of alternative relaxor ferroelectrics with high Curie temperatures and high orientation (textured), with maximized piezoelectric coefficients, by means of the CSD method. It is reported by Sabolsky et al. that the texture ferroelectric PMNT ceramics could reach the similar dielectric and piezoelectric properties compared to ferroelectric PMNT and BSPT crystals [31-33], and Zhao et al. had reported that a novel sol-gel method had been used to synthesize nanosized BS-PT ceramic powders [25].

We adopted an modified chemical solution deposition method to prepare High-Tc and texture BSPT thin films and ceramics. Compared to other film deposition methods, such as the PLD method, this novel technique has the advantage of easy control of the stoichiometry (fundamental for ferroelectric related functional oxides), high chemical purity, high chemical and micro-structural homogeneity, possibility of large area deposition, low equipment cost, and a uniformity of the thickness of the deposited films. Furthermore, certain organic solvents were utilized to dissolve and dilute the precursors, which avoids the hydrolyzing influence in the precursor solutions preparation and the bad effect of water on the film quality during the aqueous sol-gel fabrication process.

2. Objectives

The main objective of this work is to develop a reliable processing method to obtain alternative relaxor ferroelectrics with high Curie temperature and high orientation of PINT, PYNT, and BSPT systems in forms of films. To achieve this task, we selected the BS-PT system to serve as the representative materials, and adopted the chemical solution deposition (CSD) method to growthin films. Further investigations on applying different seeds for controlling the preferred orientation in films and comparing the temperature influence and template effect in different forms with the same composition have been also carried out.

3. Preliminary Experiments

a. Preparing Precursor Solutions by CSD Method

A quantity of $Pb(Ac)_2$ and $Bi(Ac)_3$ were weighed according to the compositional ratio and put into HAc solution (100%) to dissolve. After they had been completely dissolved, 1, 2-propanediol was added to make the Pb and Bi precursor solution stable. Some of Sc_2O_3 were weighed and added into hot HNO_3 solution(96%) to prepare a Sc precursor solution. By adding $Ti(i\text{-}Pr)_4$ into the HAcAc solution, the Ti precursor was prepared. According to compositional ratios of BS-PT, we mixed the above precursor solutions of Pb, Bi, Sc, and Ti into one entity. After it was diluted with 1, 2 propanediol, 0.1M and 0.3M BS-PT precursor solutions with various composition ratios were obtained. The flow chart is shown in Figure 6. 1. The raw materials are provided by the Sigma company.

b. Preparing Thin Films by Spin Coating Method

The spin coating technique was used to prepare BS-PT thin films. The substrates are of Pt/Ti/SiO2/Si. The spin speed varied from 3000 rpm to 6000rpm and the spin time varied from 20 seconds to 60 seconds. The films were first dried on a hot-stage at a temperature from 100 to 300°C.

c. Pyrolyzing and Crystallization

After coating, the films were put into a furnace to pyrolyze and crystallize. The pyrolyzing temperature is above 450°C and the crystallization temperature varies from 500°C to 850°C with the annealing time varying from 30 minutes to 1 hour.

d. Characterization

The XRD technique was used to determine the phase of the BS-PT films. The angle of X-ray scanning ranged from 15° to 80°. The surface and cross-section morphology were studied by using the scanning electronic microscopy (SEM) method. Dielectric measurements of the films were also performed using a HP impedance/gain-phase analyzer.

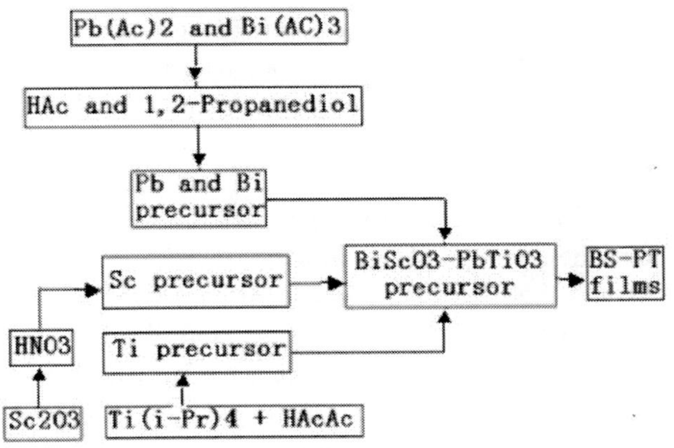

Figure 6.1. Flow chart of sample preparation.

Table 6.1. The effects of spin speed and time on the growth of BS-PT(50/50) films

Speed I (rpm)	Time I (sec)	Speed II (rpm)	Time II (sec)	Result (film quality)
0	0	3000	30	Not good
0	0	3000	60	Not good
1800	10-15	3500-4500	30-50	Good
1800	10-15	6000-7000	60	better

4. Preliminary Results

a. Thin Film Preparation

The first difficulty in preparing BS-PT films is to prepare stable BS-PT precursors. As shown in Figure 6. 1, Pb, Bi, and Ti precursors can be easily formed by dissolving $Pb(Ac)_2$, $Bi(Ac)_3$, and $Ti(i\text{-}Pr)_4$ into HAc and HAcAc solutions, respectively. But it was found that the Sc precursor was not easy to prepare using Sc_2O_3 powders. To resolve this problem, we adopted the approach of adding the Sc_2O_3 powders into dilute HNO_3 solution (96%), and heating it to 120°C for about 2 hours to dissolve, then mixing with HAc solution and 1, 2-Propanediol, and eventually, a stable Sc precursor was successfully prepared. By mixing the above four precursor solutions it was possible to prepare very stable BS-PT precursor solutions.

We prepared BS-PT thin films by employing the spin coating method. One of the difficulties in this process is that the viscosity of the BS-PT precursors is very high, and their wettability is not good enough for spanning the solution on the substrate of $Pt/Ti/SiO_2/Si$ with the spin speeds reaching 3000rpm, even if the spin time is prolonged to 60 seconds. To overcome this problem, we increased the spin speed to about 6000rpm, and found that the solution can be deposited on the substrate surface, but shrinks quickly. So, when a film is prepared, it is necessary to immediately put the samples on the surface of a hot-stage to dry. The drying temperature is 100-300°C. This process evidently affects the properties of the films. Table.1 shows the effects of spin speed and time on the growth of BS-PT(50/50) films.

BS-PT films with various compositions were put into the furnace to pyrolyze and crystallize. In our experiments, the pyrolyzing temperature and time duration is 450°C and 60 minutes, and the crystallization temperature ranges from 500°C to 800°C. The hold time duration is 30 minutes.

b. Effects of Crystalline Temperature on Phase Formation

It is crucial to determine the appropriate temperature of forming the pure perovskite BS-PT phase. In order to study the phase formation of BS-PT films under different conditions, several samples under the same pyrolyze temperature and different crystalline temperatures were prepared, and the X-ray diffraction method were performed. Figure 6. 2 shows the XRD curves of BS-PT(50/50) films on Pt/Ti.SiO2/Si substrates under different crystalline temperatures (varying from 500°C to 800°C, with a hold time of 30 minutes). Comparing these obtained XRD patterns with XRD PDF card data of TiO_2, PbO, $PbTiO_3$, Bi_2O_3, Sc_2O_3, and substrates, and with the XRD patterns of BS-PT reported by others, it is concluded that between 600°C and 800°C, BS-PT systems can achieve a pure perovskite phase. However, at 500°C, it is found that the main phase is perovskite BS-PT, but mixed with a small quantity of other phase or materials, such as Bi_2O_3 and PbO.

c. Effect of Temperature on the Development of Film Texture

After obtaining the pure phase, we focused our attention on improving the texture degree of the films. For the XRD of BS-PT(50/50) obtained from 600-800°C, as can be seen in Figure 6. 2, the intensity maximum appears at the (100) peak, showing a (100)-oriented film texture. In addition, one considerably small hump only appears at 800°C. It is also noted that the intensity of (110) peak varies by increasing the crystalline temperature. However, we cannot draw a conclusion that the films prefer to be of (110) axial orientation, since the (110) peak is the most strong among the peaks of the BS-PT system. It was found that BS-PT(50/50) thin film can achieve (100) orientation at 600°C, 650°C, 700°C, 750°Cand 800°C, respectively. Table. 6. 2 shows the effects of crystallization temperature on the phase formation and possible orientation of BS-PT(50/50) films.

In order to indicate the optimal crystalline temperature for obtaining the strongest (100)-orientation, the diffraction intensity, FWHM, and relative intensity of (100) peaks to (110) peaks under different temperature are given, as shown in Figure 6. 3. The relative intensity of (100) peak to (110) peak is obtained from the formula as following:

$$I_{(100)relative} = I_{(100)} / [I_{(100)} + I_{(110)}],$$

where $I_{(hkl)}$ stands for the intensity of (hkl) peaks in θ-2θ scan.

As can be seen in Figure 6.3 (a), the intensity of (100) peak increases with elevating temperature from 600°C to 700°C, but begins to decrease with further increases in the crystalline temperature. The trend of the FWHM value of (100) peak is not identical as that of intensity, and the minimum of FWHM appears at 700°C. In addition, the relative intensity of (100) peak shows a maximum at 700°C, as shown in Figure 6. 3(b). It is concluded that the crystallinity degree and (100) orientation degree obtained at this temperature point is the best. These phenomena indicate that, the optimal crystalline temperature on the (100)-oriented texture formation is 700°C.

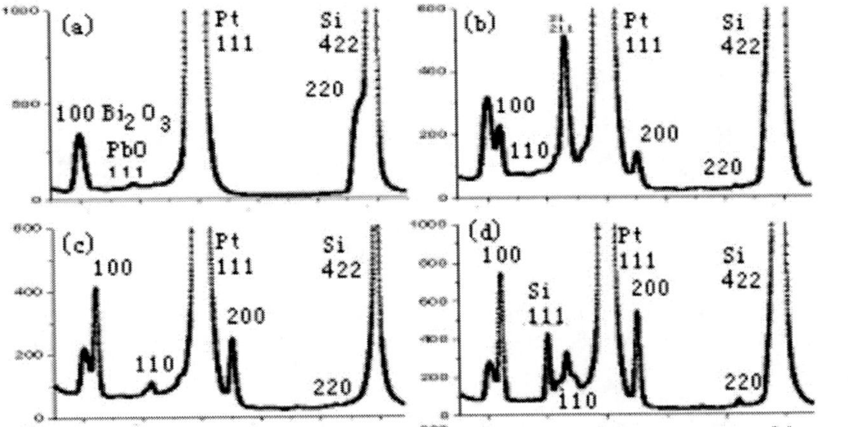

Figure 6.2. the XRD curves of BS-PT(50/50) films on $Pt/Ti/SiO_2/Si$ substrates under different crystalline temperature (a. 500°C, b. 600 °C, c. 650 °C , d. 700 °C, e. 750 °C, f. 800°C).

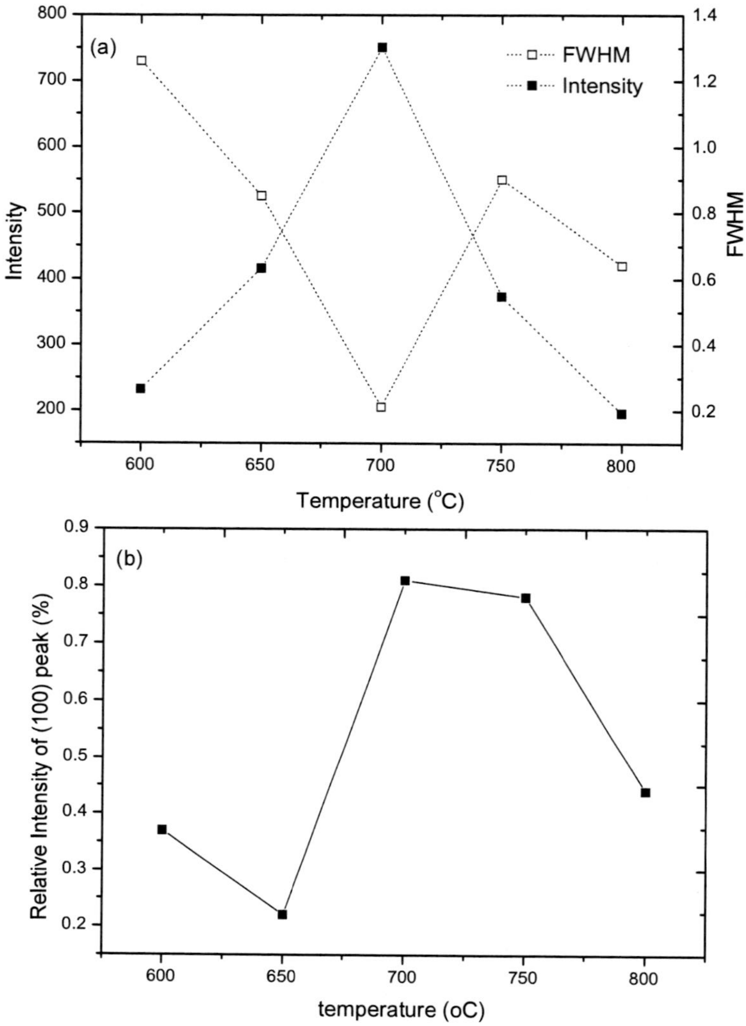

Figure 6.3. (a) relationship between the intensity and FWHM of (100) peak and the crystalline temperature; (b) relationship between the relative intensity of (100) peak and the crystalline temperature.

Table 6.2. The effects of crystallization temperature on the phase formation and possible orientation of BS-PT(50/50) films

Crystalline temperature (°C)	500	600	650	700	750	800
Pyrolyze temperature (°C)	450	450	450	450	450	450
Phase Orientation	Perovskite and other phase	Pure perovskite 100	Pure perovskite 100	Pure perovskite 100	Pure perovskite 100	Pure perovskite 100

This result is in agreement with those previously reported [30, 33]. Yoshimura et al. found that, by employing the pulsed laser deposition method, pure perovskite BS-PT(50/50) thin films with (001)-orientation were able to be obtained in a range of substrate temperature (650-730°C), and the thin films deposited at 700°C had the smallest FWHM values. Wen and Wang et al. also reported that (100)-orientated BS-PT(34/66) thin films synthesized via an aqueous sol-gel method at 700°C showed a good crystalline quality. The effect of crystalline temperature on the crystalline and texture degree of BS-PT thin films prepared by CSD method will be further investigated and discussed later.

d. Effect of Seed Layers on the Texture Formation

It was reported that, by introducing a seed layer of $PbTiO_3$, PbO, or TiO_2, the crystallization temperature of PZT films could be lowered, and the texture degree of PZT films could be improved [32]. In order to increase the texture degree and lower the crystallization temperature of BS-PT films, we also introduced a seed layer of $PbTiO_3$, PbO, or TiO_2, respectively. The XRD curves shown in Figure 6.4 show the primary results. By introducing a $PbTiO_3$ seed layer, as shown in Figure 4(a), it seems that BS-PT films formed at 650°C is of pure perovskite phase, indicating that the crystalline temperature is lower than that without a seed layer. The film is of still random and the texture is not distinct.

Distinctly, the (110) peak, which appearing in Figure 6.4(a), disappears in Figure 6.4 (b) and (c), which seems to indicate that the films with a PbO or TiO_2 seed layer show considerably strong texture degree. It is clear that the BS-PT films with a PbO or TiO_2 seed layer prefer the (100) orientation, as shown in Figure 6.4 (b) and (c). Furthermore, it is found that (100)-oriented texture degree of BS-PT films with a PbO seed layer is considerably stronger than that with a TiO_2 or $PbTiO_3$ seed layer. Table. 6.3 shows the effect of different seed layers on the crystalline and texture degree of BS-PT films.

In addition to the effect of Lattice match between the PbO seed layer and BS-PT thin films, this preferential (100)-orientation in BS-PT thin films with a PbO seed layer is also possibly due to the excess lead addition to the precursor solution introduced by a PbO seed layer, and the excess lead forms the transient phase of (100) textured PbO at the interface between substrate and BS-PT films, which results in the (100) orientation of the BS-PT films. PbO is a very good candidate for serving as a seed layer to prepare high texture BS-PT thin films. This issue will be investigated in detail later and in the next stage of this project.

Table 6.3. Effect of different seed layers on the crystalline and texture degree of BS-PT films

Seed layer	Orientation peak	Intensity (a.u)	Relative Intensity of (100) (%)	FWHM(o)
PT	100	136	53	0.8
PbO	100	2888	≈100	0.15
TiO2	100	348	≈100	0.7

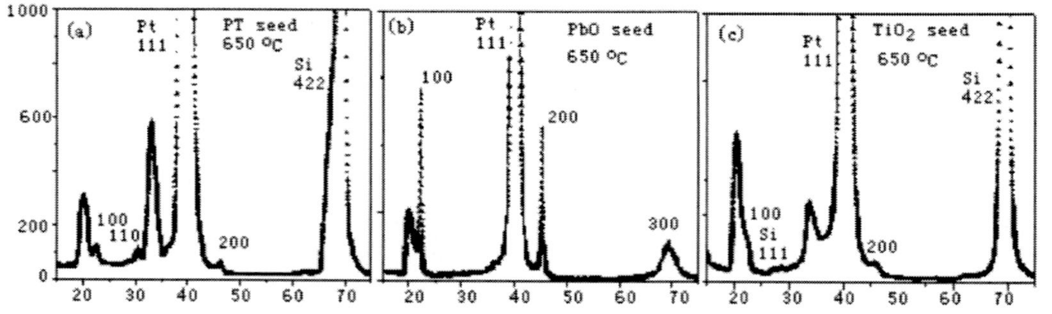

Figure 6.4. XRD pattern of BS-PT(50/50) thin film prepared by introducing a $PbTiO_3$ (a), PbO (b) or TiO_2 (c) seed layer.

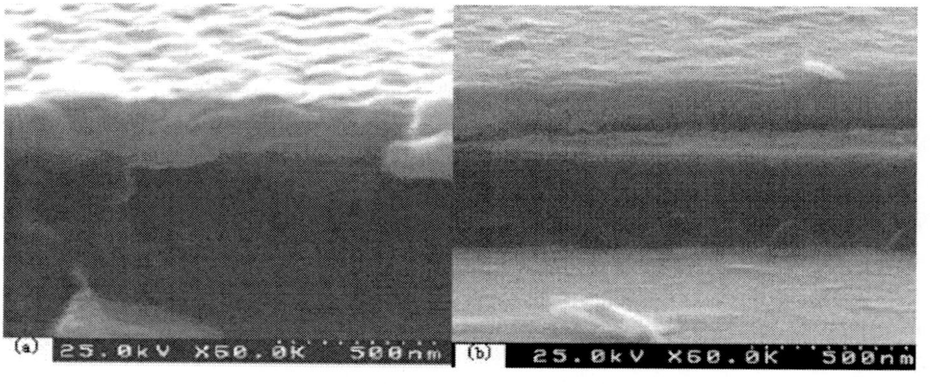

Figure 6.5. SEM morphology of BS-PT (50/50) films on different kind of seed layers.(a)on PbO seed layer;(b))on $PbTiO_3$ seed layer.

Figure 6.5 shows the SEM morphology of BS-PT(50/50) thin films with different seed layers. It is observed that the surface morphology of BS-PT(50/50) thin films with a PbO and $PbTiO_3$ seed layer show the wave topography, and the particle size on the surface is about 30nm. From the section morphology, we can estimate that the thickness of the BS-PT films is about 200-250nm.

e. Effect of Film Thickness and Rapid Annealing on the Development of Texture of BS-PT Films with and without a $PbTiO_3$ Seed Layers

To further investigate the seed layer effect on the texture, two BS-PT (36.5/63.5) films with different thickness (2 layers and 8 layers, respectively) on a PbTiO3 layer were prepared and characterized by the XRD method. The concentration of the solution is 0.3M. Special thermal processes were adopted by drying on a hot-stage at 350°C, pyrolyzing at 600°C for 20 minutes, and rapid annealing at 870°C for 30 seconds. For comparison, one 8 layer BS-PT (36.5/63.5) film without a seed layer was also prepared. The XRD results are shown in Figure 6.5, from which it can be confirmed that the films with 2 and 8 layers BS-PT films on 1 PbTiO3 seed layer have the (100) orientation, for the intensity of (100) and (200) peaks are stronger than (110) peaks, respectively. However, it is noted that, under such thermal conditions, with the thickness increases, the texture degree of the films decreases. In contrast, Figure 6.5(c) indicates that the BS-PT film without a $PbTiO_3$ seed layer is random.

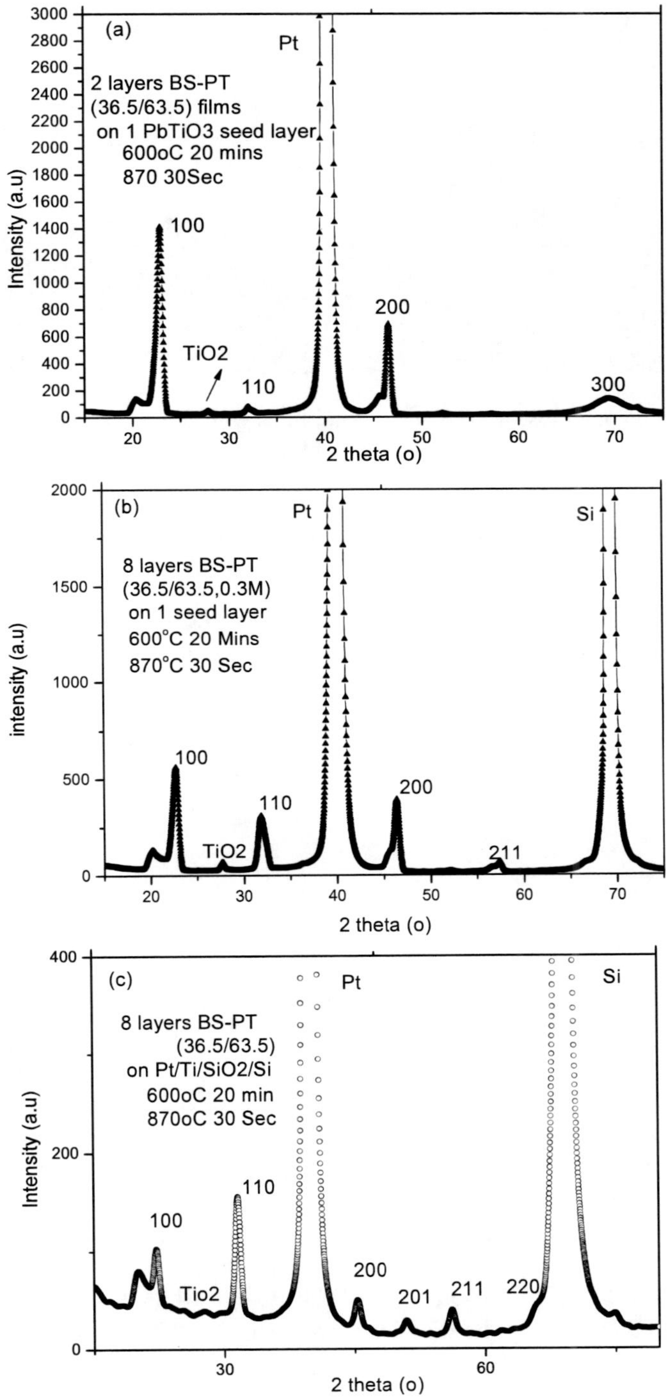

Figure 6.5. XRD patterns of 2 layers (a) and 8 layers (b)8 layers BS-PT films deposited on a PbTiO3 seed layer and $Pt/Ti/SiO_2/Si$ substrate, and 8 layers BS-PT films without PbTiO3 seed layer (c).

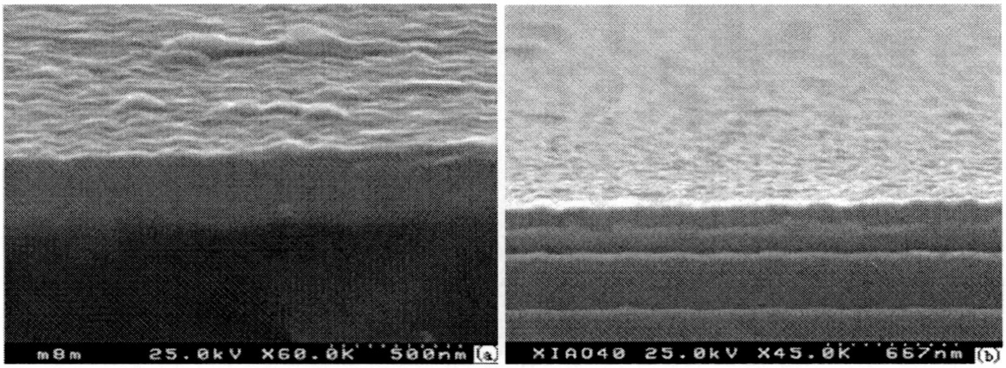

Figure 6.6. Sem morphology of BS-PT(36.5/63.5) thin films with and without a $PbTiO_3$ seed layer. (a)corresponding to Figure 9 (b);(b) corresponding to Figure 9 (c).

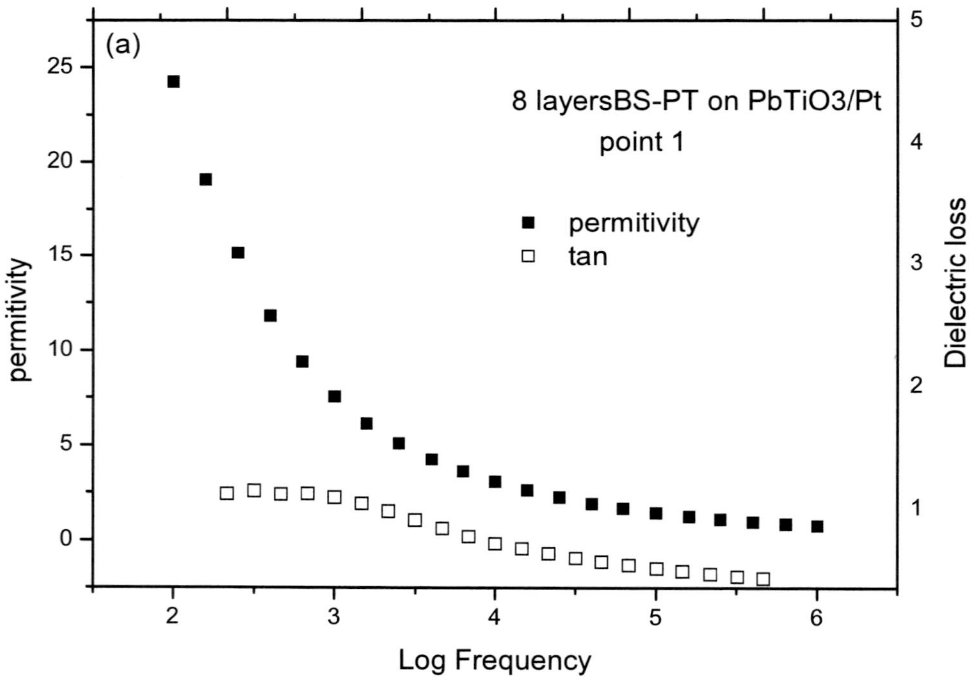

Figure 6.7. Dielectric properties of BS-PT thin film corresponding to Figure 6.9(b).

Figure 6.6 shows the SEM morphology of the BS-PT films corresponding to Figure 6.5 (b) and (c). Figure 6.7 indicates the dielectric properties of 8 layers BS-PT(36.5/63.5) films with a seed layer. Compared with what has been reported previously, the present permittivity is very low, while the dielectric loss is very high. This phenomenon is probably because they are not pure perovskite phase, for at about a 27.8^{o} angle range, there exists a small peak which corresponds to the (110) peak of TiO_2 (rutile phase). The reason for the appearance of the TiO_2 phase is the fact that, during the high temperature process, the volatilization of PbO leads to an excess Ti component, from which is easy to form a small quantity of TiO_2 rutile phase under the condition of high temperature and atmosphere.

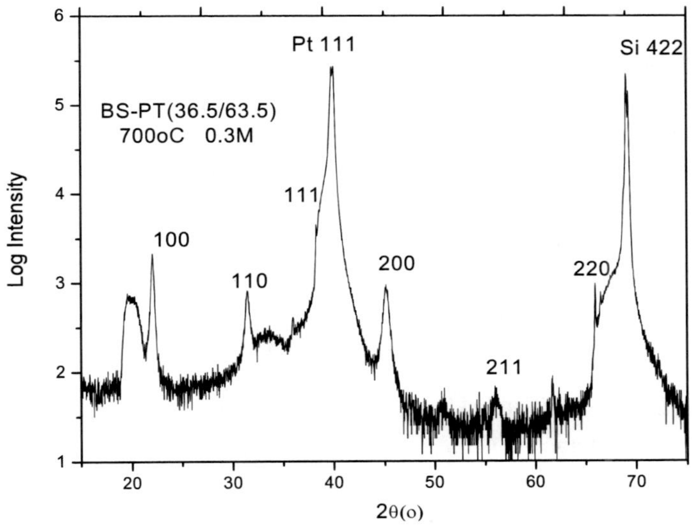

Figure 6.8. XRD curve of BS-PT(36.5/63.5) films prepared by 0.3M solution.

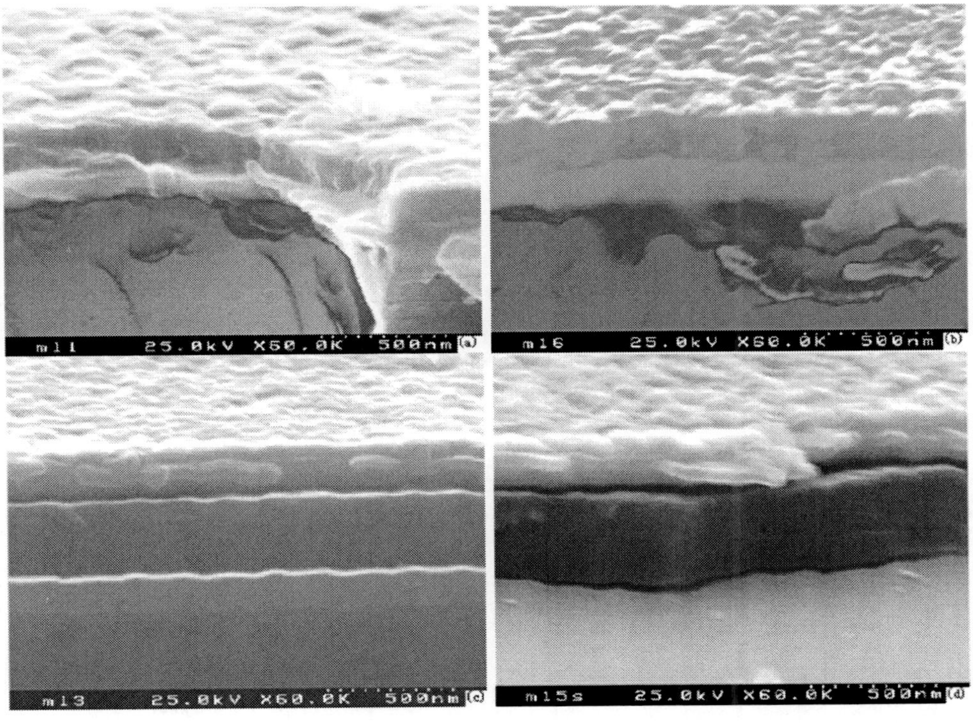

Figure 6.9. SEM morphology of different BS-PT (36.5/63.5) film layers on $Pt/Ti/SiO_2/Si$ substrate and a $PbTiO_3$ seed layer. (a)13 layers BS-PT films on a seed layer of $PbTiO_3$ on $Pt/Ti/SiO_2/Si$, the solution concentrate is 0.1M; (b)3 layers BS-PT films on a seed layer of $PbTiO_3$ on $Pt/Ti/SiO_2/Si$; the solution concentrate is 0.3M;(c)10 layers BS-PT films on $Pt/Ti/SiO_2/Si$, the solution concentrate is 0.1M; (d)10 layers BS-PT films on $Pt/Ti/SiO_2/Si$, the solution concentrate is 0.3M.

f. Effects of Solution Concentrations and Layers on the Microstructure of BS-PT Thin Films

Figure 6.8 shows an XRD curve of BS-PT(36.5/63.5) films prepared by 0.3M solution. It can be observed that they also show the perovskite phase. It is expected that the concentration of the solution plays an important role in preparing BS-PT thin films.

Figure 6.9 shows images of SEM morphology of different BS-PT (36.5/63.5) film layers on Pt/Ti/SiO_2/Si substrate and a $PbTiO_3$ seed layer, with different precursor concentrates is respectively 0.1M and 0.3M, from which it is possible to understand the effect of the concentration of solutions and film layers on the microstructure of BS-PT films with and without a $PbTiO_3$ seed layer. As shown in Figure 6.9(a) and (b), on a seed $PbTiO_3$ layer, the BS-PT film with 13 layers by a solution concentration of 0.1M seems to show the character of columnar grain growth, while the film with 3 layers by a solution concentration of 0.3M shows a grainy growth character. This phenomenon is expected to be in agreement with what had been reported by S. Hoffmann et al, which implied that $BaTiO_3$ and $SrTiO_3$ films exhibit a columnar grain growth for a small solution concentration (0.1M) and more crystallized layers. But the effect of the substrate cannot be ignored, for BS-PT films without a $PbTiO_3$, the SEM morphology exhibit grainy growth. The concentration is 0.1M and the number of layers is not less, as shown in Figure 6.9(c) and (d).

g. Effect of the Changes of Viscosity and Wettability of Solutions

In addition, in order to decrease the viscosity of these BS-PT solutions and improve the wettability between the substrate and the precursors, we respectively introduced a small quantity of Formamid and Acetone as a dilute solvent into the 0.1M BS-PT(36.5/63.5) solution. From the XRD curves shown in Figure 6.10, it is indicated that adding these two solvent is also able to obtain a pure perovskite phase. However, the effect of these reagents for preparing high quality BS-PT thin films is not distinct.

Figure 6.11 shows the SEM surface morphology of some BS-PT thin films by using Acetone and Formamide as dilute solvents. From the images, it is observed that the stack of grains is not intact since the surface is full of small holes and porous.

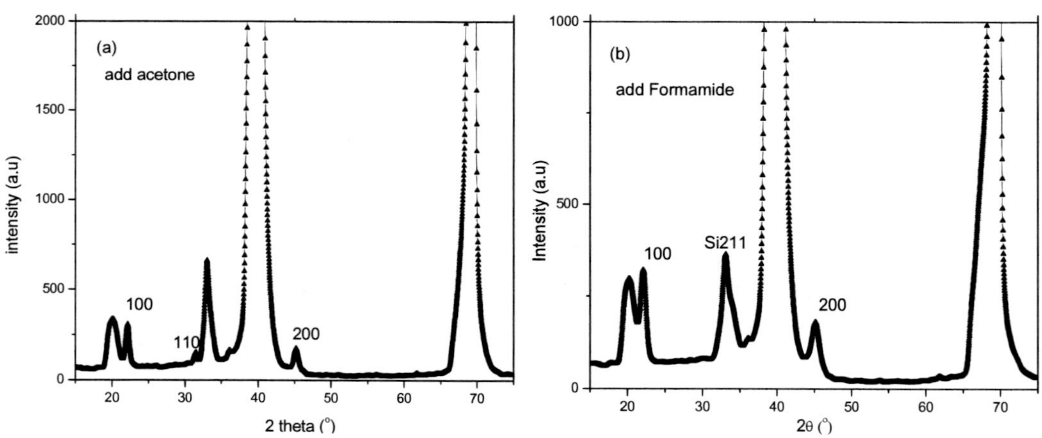

Figure 6.10. XRD curves of BS-PT(36.5/63.5) films prepared by 0.1M solution diluted by acetone (a) and formamid (b).

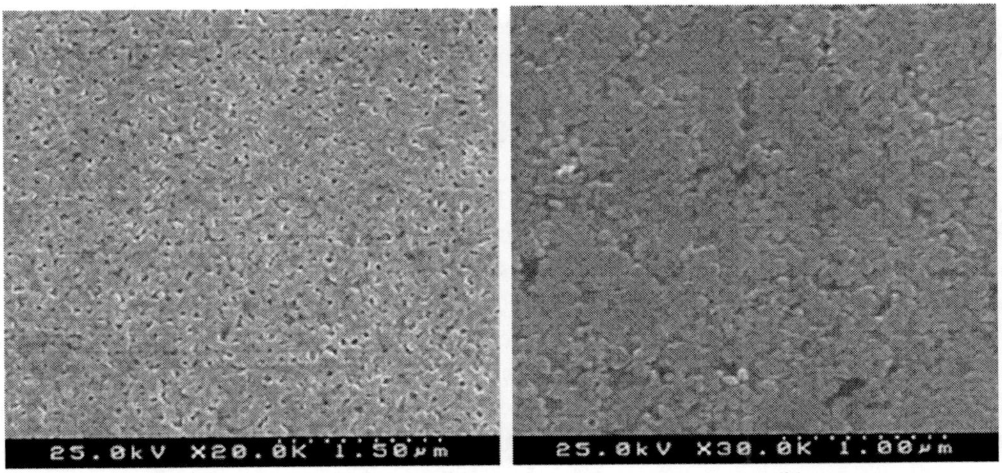

Figure 6.11. Surface SEM morphology of BS-PT thin films.(a) BS-PT(36.5/63.5), 0.1M, add acetone as solvent; (b) BS-PT(36.3/63.5), add formamid to dilute to 0.1M.

We measure the dielectric properties at room temperature of a BS-PT sample (using Acetone as dilute solvent). Compared to that value of BS-PT single crystals (about 800), it is found that the dielectric constant is also very small, and the dielectric loss is also large. The reason is considered to be due to the fact that the films are very thin or the grains are not in contact with each other.

According to the preliminary results, it is implied that applying the CSD method can successfully prepare high texture BS-PT film with high-Tc, which is expected to improve the piezoelectric properties with high-Tc.

(1) One of the difficulties in fabricating thin films by the CSD method is that of preparing stable precursor solution systems. In this annual project stage, we firstly focus our attention on preparing the BS-PT precursor solutions with different processes, and some stable BS-PT precursor solutions have been successfully obtained. It implies that we had solved the technique problems in preparing BS-PT precursor solutions. This accomplishment makes the CSD deposition of BS-PT films feasible.
(2) Applying the spin coating method, BS-PT thin films with different compositions have been prepared. By optimizing the thin film growth parameters, the effect of the spin rate and spin time have been investigated.
(3) To prepare pure perovskite phase BS-PT thin film, the post-coating thermal processes have been studied and the optimized pyrolyze and crystalline temperatures have been determined.
(4) The work for determining the thin film structures and phases of BS-PT films has been successfully carried out by the XRD method. The XRD results show that the pure perovskite phase of BS-PT films can be obtained.
(5) The surface and cross-section morphology observation on BS-PT thin films has been performed by the SEM techniques, from which the thin film's grain size and

thickness can be obtained. Electricity measurements on BS-PT thin films has been preliminarily performed.

6.2. Sol-Gel Derived MPB $0.37BiScO_3$-$0.63PbTiO_3$ Thin Films

The sol-gel derived MPB $0.37BiScO_3$-$0.63PbTiO_3$ ferroelectric thin films with a $PbTiO_3$ seed-layer deposited on platinized silicon substrates exhibit excellent dielectric/ferroelectric properties. A room temperature dielectric constant of >1500 and dielectric loss of <0.02 are achieved (100 Hz). A well-defined hysteresis loop was observed with a P_r of ~23μC/cm^2. In particular, the remarkably low E_c of ~34kV/cm of these films adds value to the potential application of $BiScO_3$-$PbTiO_3$ films in high temperature ferroelectric memories. The influence of the $PbTiO_3$ seed-layer on the electric properties and the relation with the phase formation process, crystallinity and microstructure of the films is discussed.

It is well known that perovskite ferroelectric thin films have attracted considerable attention as promising materials for utilization in memory devices, microelectromechanical (MEMs) [34-36] and more recently in nanoelectromechanical applications. [37] Possessing a high remanent polarization (P_r of ~30 μC/cm^2) and low coercive electric field (E_c of 50-70 kV/cm), $Pb(Zr_{1-x}, Ti_x)O_3$ (PZT) has become one of the most popular piezoelectric/ ferroelectrics for these applications. [38] Meanwhile, the industrial community has been expressing the need for sensing and actuation over a broad temperature range. [39] In addition to optimized electric performances, ferroelectrics with high Curie temperature (high-T_c) and good temperature stability of the properties are also required and highly desirable for special applications under high temperature environments. High-T_c ferroelectrics are desired, for example, in memory devices for planetary exploration and other space applications, which requires technology that allows storing, retrieving and processing information under harsh temperature conditions in outer space.

Recently, predicted by a perovskite crystal structure tolerance factor relationship, compositions based on $(1\text{-}x)$ $BiScO_3$-$xPbTiO_3$ (BSPT) system were recognized as a potential candidate for these applications. [39-46] According to the $(1\text{-}x)BiScO_3$-$xPbTiO_3$ phase diagram, rhombohedral and tetragonal phases and a morphotropic phase boundary (MPB) were identified within the following composition range, $0.50<x<0.62$, $x>0.65$ and $0.62<x<0.65$, respectively. It is now recognized that the BSPT system possesses important advantages over the PZT family, such as a T_c ~ 450°C, circa 100 °C higher than that of PZT and a temperature stability of the electrical properties up to 350 °C. Simultaneously, this system exhibits very good electric properties comparable to those of $PbTiO_3$-based ferroelectrics near the morphotropic phase boundary (MPB) compositions, for instance, a P_r of ~ 32 μC/cm^2 and a piezoelectric coefficient, d_{33} on the order of 500 pC/N were observed in BSPT bulk materials. All these observations have been the driving force for the continuous research on BSPT system (ceramics and crystals), since its discovery by Randall et al, in 2001. The justification for further investigations on BSPT films are also related to the current interest in thin films to meet the miniaturization requirements of the microelectronics industry.

However, only a few works on high-T_c BSPT thin films have been reported. Yoshimura et al. fabricated 1-μm thick epitaxial films of rhombohedral (001) $(1\text{-}x)$ $BiScO_3$-$xPbTiO_3$ (with x= 0.60 and 0.50) on (100) $SrRuO_3$/(100) $LaAlO_3$ substrates by pulsed laser deposition. [47, 48] A room temperature permittivity and dielectric losses of about 850 and 0.08, respectively, well-saturated hysteresis loops with remanent polarization of 36-42 μC/cm^2 and coercive field of 65-75 kV/cm were observed. H. Wen et al prepared (100)-oriented tetragonal (1-x) $BiScO_3$-$x$$PbTiO_3$ (x=66) thin films by aqueous sol-gel method. [49]

Table 6.4. Ferroelectric properties of BSPT thin films obtained at room temperature in this work and reported in the literature

$BiScO_3$-$PbTiO_3$ solid solution	Film thickness (nm)	Remnant polarization (μC/cm^2)	Coercive field (kV/cm)	Dielectric permittivity and loss	Crystal phase and references
$0.50BiScO_3$-$0.50PbTiO_3$	1000	36	65	800, 0.09	R* [14]
$0.40BiScO_3$-$0.60PbTiO_3$	1000	42	75	850, 0.08	R* [15]
$0.34BiScO_3$-$0.66PbTiO_3$	200	33	170	1200, 0.05	T* [16]
$0.34BiScO_3$-$0.66PbTiO_3$	250	50	200	1609, 0.08	T* [17]
$0.34BiScO_3$-$0.66PbTiO_3$	350	74	220	2000, 0.07	T* [18]
$0.34BiScO_3$-$0.66PbTiO_3$	600	74	177		T* [19]
$0.37BiScO_3$-$0.63PbTiO_3$	500	22.87	33.6	1531, 0.02	MPB* Present work

(*Note: R stands for rhombohedral phase, T stands for tetragonal phase, MPB stands for morphotropic phase boundary.)

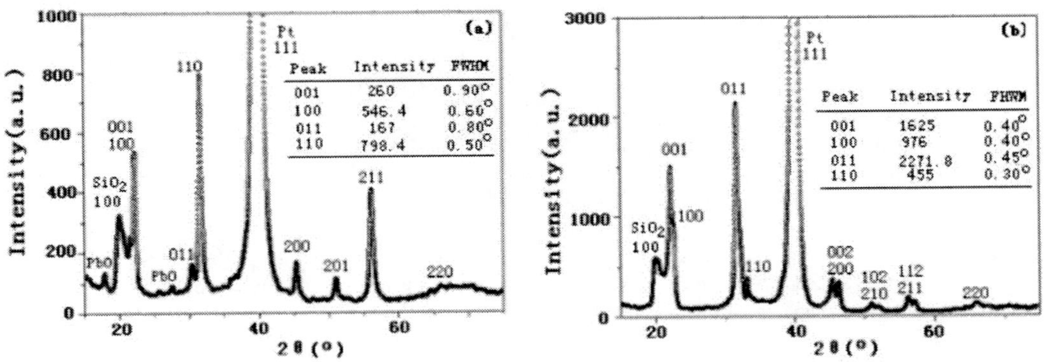

Figure 6.12. XRD patterns of BSPT (37/63) films on Pt/Ti/SiO_2/Si without (a) and with a $PbTiO_3$ seed layer (b). The inset corresponds to the values of intensity and FWHM of the main diffraction peaks of the unseeded and $PbTiO_3$ seeded BSPT (37/63) films.

Room temperature dielectric constant and dielectric losses achieved values of 1200 and 0.09, respectively. These authors also reported saturated hysteresis loops with remanent polarization of 33μC/cm^2 but with very high coercive fields of around 220 kV/cm. In their subsequent work, [50-53] variations on the ferroelectric properties of thin films with the same composition were reported, as shown in Table 6.4, in which high coercive field values were always presented.

In view of their utilization in ferroelectric memory devices, ferroelectric films with good square hysteresis loop, with high remanent polarization (P_r) and in particular low coercive fields (E_c), are required to allow operation at low voltages, short switching time, good retention and fatigue properties. However, based on the reported results, both the rhombohedral and tetragonal BSPT films exhibit too high coercive fields, severely restricting their potential use. As a consequence, it is of significant interest to fabricate high-T_c BSPT ferroelectric films with low coercive fields.

Figure 6.13. Top view ((a) and (c)) and cross-section ((b) and (d)) SEM images of 10 layers BSPT (37/63) films on $Pt/Ti/SiO_2/Si$ without ((a) and (b)) and with a $PbTiO_3$ seed layer ((c) and (d)). (e) Cross-section SEM image of a $PbTiO_3$ seed layer on Pt electrode; the seed layer thickness is about 60 nm.

It was reported that, in PZT and other lead-based ferroelectrics, [54-56] compositions located at MPB exhibit minimum coercive fields, which have been attributed to the increased number of possible crystallography and spontaneous polarization directions, due to the coexistence of rhombohedral and tetragonal phases. Supporting this observation a minimum coercive field of ~20 kV/cm was reported in MPB $0.37BiScO_3$-$0.63PbTiO_3$ ceramics. [57] Therefore, analogously, $0.37BiScO_3$-$0.63PbTiO_3$ thin films with MPB composition are anticipated to exhibit lower coercive fields when compared to the rhombohedral and tetragonal BSPT films. Nevertheless, no reports on the research of MPB BSPT thin films have been found.

In this work, MPB $0.37BiScO_3$-$0.63PbTiO_3$ thin films (here abbreviated as BSPT (37/63)) are fabricated by the sol-gel method, and the ferroelectric properties, namely the ferroelectric switching behavior evaluated.

The precursor solution with the starting composition of BSPT (37/63) and concentration of 0.2M was synthesized by a sol-gel approach similar to that reported by Wen et al. for BSPT (34/66). But in contrast, MPB BSPT solutions were found to be quite difficult to be deposited on bare $Pt/Ti/SiO_2/Si$ substrates, due to the poor adhesion of the film to Pt and also to densification difficulties. Because of that, seed layers of $PbTiO_3$ were introduced, to improve the solution adhesion to the substrate and the crystallization, and then to enhance the electrical response. The process of depositing the $PbTiO_3$ seed layers on $Pt/Ti/SiO_2/Si$ substrates is similar to that reported by Zeng et al. [58] The BSPT (37/63) films were deposited by spin coating the solution on the top of $Pt/Ti/SiO_2/Si$ substrates with and without a seed layer at 3000-5000 rpm for 30 seconds. After being dried at 350 °C and pyrolyzed at 650°C for several minutes, the films with the desired thickness were heat treated at 750°C for 2 hours using a conventional thermal annealing (CTA) process to crystallize the peroveskite BSPT (37/63) phase and to densify the films. In this annealing process, the film samples were placed into the PbO and Bi_2O_3 atmosphere to depress the possible volatilization of these two compositions. For comparison, 10 layer BSPT (37/63) thin films with and without $PbTiO_3$ seed layers were prepared under the same conditions.

XRD and scanning electron microscopy (SEM) techniques were used to identify the crystalline phases that were formed and to analyze film microstructures. For electrical measurements, gold top electrodes with a diameter of 0.6mm were deposited on the films by DC sputtering using a shadow mask. The dielectric and ferroelectric characterization was performed at room temperature using an HP4284 Precision LCR Meter and TF analyzer (AIXACCT, Model TF A-LI), respectively.

The XRD patterns of BSPT (37/63) thin films with and without $PbTiO_3$ seed layer are depicted in Figure 6.12. Unseeded BSPT film exhibits the MPB perovskite phase with a random orientation. Some extra peaks of PbO second phases are also observed at 27.8° and 25.5° (2θ). The film with a $PbTiO_3$ seed layer shows a pure MPB BSPT (37/63) phase, with also no preferred orientation. As shown in the inset tables of Figure 6.12, the intensities of (100) and (110) peaks of the seeded BSPT(37/63) films increased, while the values of the full width at half maximum (FWHM) decreased, which clearly demonstrates that seed layers enhanced the crystallinity of BSPT thin films.

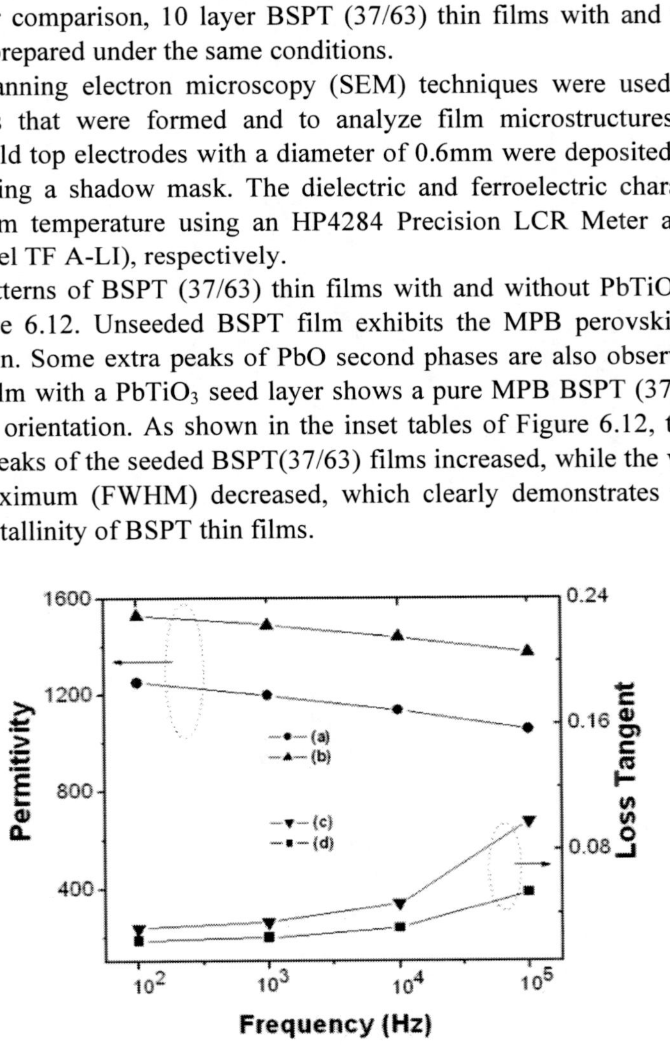

Figure 6.14. Dielectric permittivity as a function of frequency for unseeded (a) and seeded (b) BSPT (37/63) films and dielectric loss as a function of frequency for unseeded (c) and seeded (d) BSPT (37/63) films.

Figure 6.13 depicts the SEM microstructures of dense unseeded and seeded BSPT (37/63) films. The film thickness and average grain size of seeded films are about 500 and 150 nm, while those of unseeded ones are about 300 and 120 nm, respectively. Seeded films present slightly higher thickness and bigger grains than unseeded films. The improved adhesion of BSPT film to the substrate due to the $PbTiO_3$ seed layer somehow responds to the observed increasing of seeded film thickness. In addition, the $PbTiO_3$ seed layer promotes the perovskite phase formation and enhanced the crystallization process, in accordance with the decrease of FWHM XRD peaks observed for seeded films. These effects, also observed for other perovskite thin films [59] are clearly explained from a kinetic point of view. The thin crystalline $PbTiO_3$ seed layer deposited beneath BSPT films constitutes a multitude of nucleation sites at the film/substrate interface that heterogeneously decrease the energy barrier for BSPT crystallization. As a consequence, seeded films possess the same annealing temperature at a higher degree of crystallinity than unseeded films.

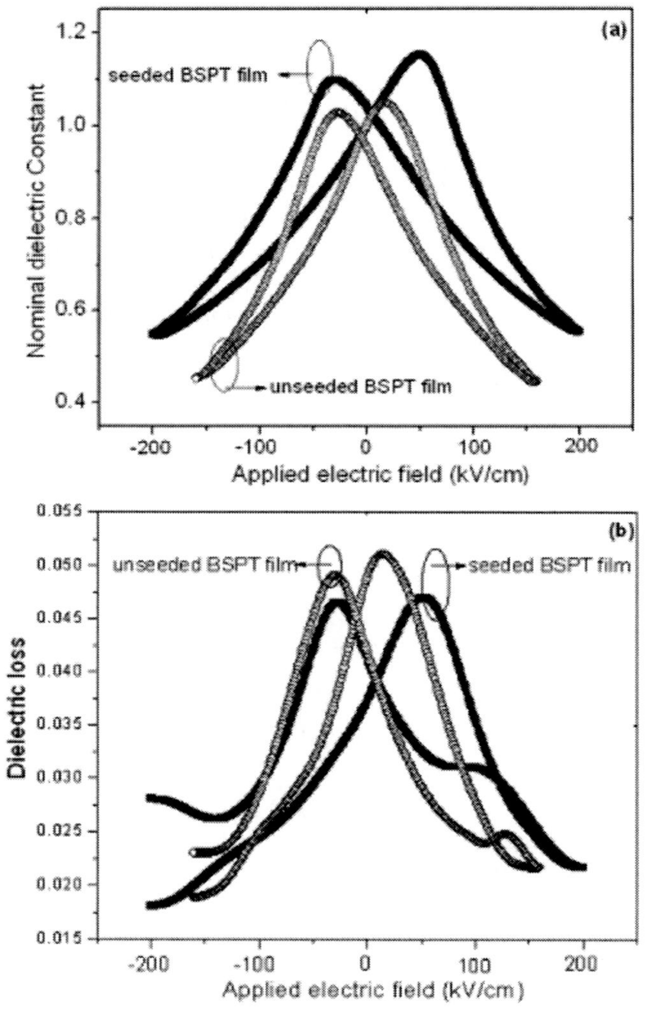

Figure 6.15. The dc bias field dependence of nominal dielectric constant ($\varepsilon/\varepsilon_{zero\ bias}$) (a) and dielectric loss (b) in unseeded and seeded BSPT (37/63) films.

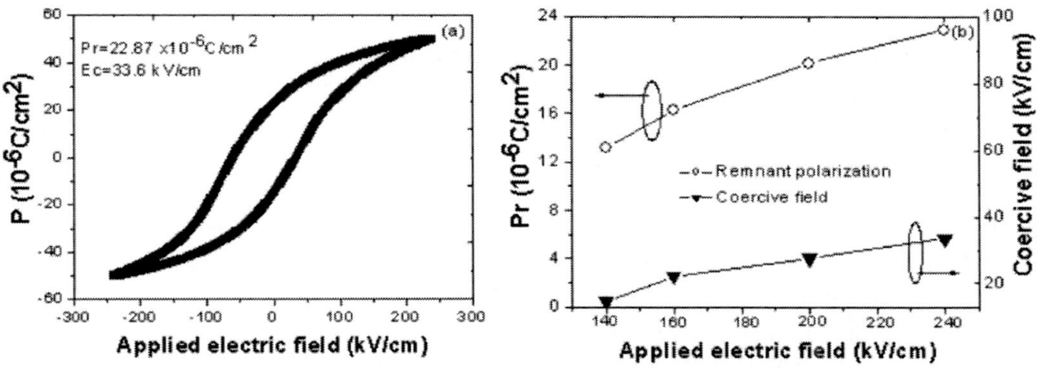

Figure 6.16. (a) Hysteresis loop of BSPT (37/63) films with a $PbTiO_3$ seed layer on Pt/Ti/SiO_2/Si; (b)Variation of remanent polarization and coercive field versus applied electric field.

Figure 6.14 illustrates the frequency dependence of dielectric constant and dielectric losses of MPB BSPT (37/63) thin films with and without $PbTiO_3$ seed layer. At room temperature, the seeded BSPT films display a dielectric constant of 1531 and dielectric loss of 0.02 at 100 Hz. As expected, seeded films exhibit higher dielectric permittivity and lower dielectric losses than unseeded films. The $PbTiO_3$ seed layer has also a positive impact on the dielectric properties, which can be ascribed to the increased degree of crystallinity of the perovskite phase and the quality of the film's microstructure. Figure 6.15 (a) depicts the dc bias dependence of a nominal dielectric constant ($\varepsilon/\varepsilon_{zero\ bias}$) for seeded and unseeded BSPT(37/63) films, which clearly illustrates that the dielectric constant of seeded films exhibits a higher increase rate in the small bias field region and a lower decrease rate in the large dc bias region. This, together with the dc bias dependence of dielectric loss of these two samples, show that the seeded BSPT(37/63) films exhibit a better dielectric response under dc bias field than unseeded ones.

Figure 6.16 (a) shows the hysteresis loop of BSPT films with a $PbTiO_3$ seed layer. Very well defined hysteresis loops were observed with a remnant polarization (P_r) of about 23μC/cm^2 and a coercive field (E_c) of circa 34kV/cm. The variation of remanent polarization and coercive field versus the applied electric field are shown in Figure 5(b), which illustrates a nearly linear variation. Furthermore, as indicated in Table 6.4, MPB BSPT (37/63) thin films prepared in this work show a remnant polarization value which is comparable to those reported by others, whereas it exhibits a considerably lower coercive electric field. These results highlight the merit of low coercive field values of BSPT (37/63) films that constitutes an important benefit for low voltage operation. MPB BSPT films are then more attractive for high temperature memory device applications, in comparison to previously reported rhombohedral and tetragonal BSPT films. Compared to conventional PZT films (Table 6.4) deposited on platinum substrates, BSPT(37/63) thin films also possess better ferroelectric properties.

Coercive field value is an indication of the ease of domain switching and domain polarization reversal under an electric field. In MPB compositions, in which the number of polarization directions increases due to the coexistence of rhombohedral (8 possible polarization directions) and tetragonal phases (6 possible polarization directions), the polarization reversal will be easier than in compositions out of the MPB region. As indicated in a systematic study on composition dependence of ferroelectric properties by Oikawa et al.,

[56] E_c for $Pb(Zr_x, Ti_{1-x})O_3$ thin films varied as x increased and reached the minimum (about 50 kV/cm) at MPB PZT (52/48) composition. A similar tendency was observed in this work in which BSPT(37/63) thin films within the MPB composition range exhibit considerably lower coercive field values than equivalent sol gel BSPT films with compositions out of the MPB range. E_c of 65 kV/cm and 220 kV/cm were reported for BSPT(50/50) and for BSPT(34/66).

However coercive field values are also dependent on many external parameters including thermal and electrical history, environmental factors, as well as the type and area of switching electrodes. Microstructure features will contribute extrinsically to the domain switching kinetics, being one of the most important, in polycrystalline films which is the grain boundary pinning effect. [60] As the grain size increases, a decrease of the coercive field is expected, as observed for BSPT seeded films and also reported for several systems.

The other aspect to take into account is that seed layers affect markedly the film defect state, mainly at the interface of the film/substrate. The $PbTiO_3$ seed layer will decrease the concentration of interfacial defects between BSPT and the Pt layer that may also contribute to the decreased pinning points or domain switching, decreasing E_c. However this hypothesis deserves further analysis. The importance of the low E_c values obtained in this work is further reinforced if one considers that the coercive field (E_c) of ferroelectric films decreases with the increase of film thickness. The thickness of the films prepared in this work (500 μm) is half of the thickness of BSPT films prepared by Yoshmura et al.

In summary, MPB $0.37BiScO_3$-$0.63PbTiO_3$ thin films were fabricated by the sol-gel method with a $PbTiO_3$ seed layer, which exhibit a pure perovskite phase and improved crystallinity. Good electric properties with a dielectric constant of >1500, dielectric loss of 0.02, and remanent polarization of 22.9 μC/cm^2 were observed. A very low coercive field value of 33.6 kV/cm shows that the MPB $BiScO_3$-$PbTiO_3$ thin films have potential applications in high temperature ferroelectric memory devices, and will stimulate further exploration in high-Tc ferroelectric films.

6.3. MPB $0.37BiScO_3$-$0.63PbTiO_3$ Films Deposited on Iridium Oxide Electrodes

To identify the influence of the IrO_2 electrodes on the film's microstructure and electrical properties, the characterization of sol-gel derived morphotropic phase boundary (MPB) $0.37BiScO_3$-$0.63PbTiO_3$ (BSPT) thin films deposited on $IrO_2/TiO_2/SiO_2/Si$ and $Pt/TiO_2/SiO_2/Si$ substrates was performed. Although the ferroelectric behavior of both films is similar, with remanent polarization values of 26 μC cm^{-2} and 23 μC cm^{-2} respectively for IrO_2 and Pt bottom electroded films, the leakage current density (J_L) at room temperature markedly decreases from > 10^{-6} A cm^{-2} for BSPT on Pt to the order of ≤ 10^{-8} A cm^{-2} for BSPT on IrO_2, under the maximum voltage of 4V (~80 kV cm^{-1}). The formation of an interface layer between the film and the electrode, as observed by Rutherford Backscattering Spectroscopy, is proposed to have a space charge sink effect and, as a consequence, to account for the improvement of the leakage current behavior of the IrO_2 electroded films.

Concerning device application, the lifetime of a ferroelectric film capacitor, or its performance stability during operation and over time, is crucial, and depends on key aspects, such as fatigue, retention time of the ferroelectric properties and leakage current, generally termed reliability. It is well known that ferroelectric thin films, such as PZT fabricated on $Pt/TiO_2/SiO_2/Si$ substrates, have serious reliability problems, in part caused by the reactions (physical and chemical) occurring between the underneath layers (electrode / substrate) and the ferroelectric film, which impairs their performance for utilization. [62-65] Some of these limitations were bypassed, in the case of PZT films, through the use of alternative oxide electrodes. These oxide electrodes act not only as a good barrier against layer inter-diffusion, but also as a "sink for defects", [64] or an entrapment of defects at the film / electrode interface, improving the reliability of the capacitor.

Two main categories of oxide electrodes have been used in these capacitors: the common metal oxides, such as IrO_2, [63-65] and RuO_2 [66] and the conductive perovskite oxides, such as $SrRuO_3$ [67] and $LaNiO_3$. [68, 69] Among the perovskite oxide electrodes, $LaNiO_3$ offers a close thermal expansion coefficient match and a good adhesion to the PZT films. The use of $LaNiO_3$ as the bottom electrode significantly improves the leakage and the fatigue behavior of PZT films. After 10^9 bipolar switching cycles, no significant change in the remnant polarization was observed for PZT/$LaNiO_3$ based capacitors, in comparison to the PZT/Pt system, which lost more than 50% of its original remnant polarization; PZT/$LaNiO_3$ films become fatigue-free for up to 10^{11} cycles. [70] PZT (20/80) thin films on $LaNiO_3$ exhibit a leakage current density of 1.38×10^{-6} A cm^{-2} at 5 V, lower than that of PZT films deposited on Pt electrodes reported by Kang *et al.* [69]

PZT films grown on IrO_2 bottom electrodes were also reported to exhibit improved reliability. Despite the slight decrease of the remanent polarization of PZT / IrO_2 (from 12-14 $\mu C\ cm^{-2}$ on Pt to 8-10$\mu C\ cm^{-2}$ on IrO_2), it remains nearly unvaried after the application of 10^{12} switching pulses when compared to a reduction of half of the remanent polarization after about 10^8 switching cycles for the films deposited on Pt electrodes, as reported by Nakamura *et al.* [63] Moreover, the use of IrO_2 as the electrode also caused a reduction of the capacitor's leakage. Leakage currents density of 6×10^{-7}A cm^{-2} and 1×10^{-6} A cm^{-2} were reported for PZT films with IrO_2 electrode and Pt electrodes, respectively.

In comparison with other conductive metal oxides (such as $LaNiO_3$), IrO_2 can be easily and reproducibly deposited on Si substrates on a large scale, keeping the costs low and meeting the integration requirements for Si-based devices. These advantages make IrO_2 electrodes very attractive for the preparation of ferroelectric thin films and related memory chips. [71-73] However, up to now, the dielectric and ferroelectric behavior of BSPT thin films deposited on IrO_2 electrodes have not been reported. The influence of $SrRuO_3$ or Nb: $SrTiO_3$ electrodes on the electric performance and reliability of the ferroelectric properties of BSPT films has also not yet been reported. In the case of Nb: $SrTiO_3$ electrodes, high costs severely restrict their broad application.

It is therefore interesting and opportune to investigate the effect of IrO_2 electrodes on the growth and electric properties of MPB BSPT thin films. In this work, MPB $0.37BiScO_3$-$0.63PbTiO_3$ films (here abbreviated as BSPT) were sol-gel deposited on $IrO_2/TiO_2/SiO_2/Si$ substrates. In comparison with the Pt based BSPT capacitors, the influences of IrO_2 electrodes and the film / electrode interfaces on improving the dielectric / ferroelectric properties and leakage current behavior were explored and discussed.

Lead acetate ($Pb(COOCH_3)_2$) (Panreac Company, 99.5%), bismuth nitrate ($Bi(NO_3)_3$) Sigma-Aldrich, 99.99+%), scandium acetate ($Sc(COOCH_3)_3$ (Sigma-Aldrich, 99.9%), and titanium isopropoxide ($Ti(OiPr)_4$) (Sigma-Aldrich, >97%) were used as starting reagents; the selected solvents were glacial acetic acid (Pronalab, 100%), 2-propanol solutions (Fluka, 99.8%), distilled water and acetylacetone (Merck, 99.5%). Formamide (Merck, 99.5%) was used as an additive to improve the adhesion of the film to the substrate. A 0.2M BSPT (37/63) precursor solution was synthesized. Commercial $IrO_2/TiO_2/SiO_2/Si$ wafers (150nm/20nm/300nm/Si substrate, Inostek Inc.) were used as substrates. Based on the previous work, 62 BSPT films with a $PbTiO_3$ seed layer intercalated between the BSPT film and the substrate were spin coated on $IrO_2/TiO_2/SiO_2/Si$ (hereafter abbreviated as BSPT/IrO_2), with a spin rate of 3000-5000 rpm and a spin time of 30 seconds. Each layer was dried at 350 °C and pyrolyzed at 650 °C for 5 minutes. After all 10 layers had been deposited, the ferroelectric films with the desired thickness (~550 nm) were heat treated at 750 °C for 2 hours. In this annealing process, film samples were placed into a Pb and Bi rich atmosphere (created by surrounding them with PbO and Bi_2O_3 powders) to suppress the possible volatilization of these elements. For comparison, BSPT thin films on $Pt/TiO_2/SiO_2/Si$ (150nm/20nm/300nm/Si substrate) with $PbTiO_3$ seed layers (hereafter abbreviated as BSPT/Pt) were prepared under identical conditions.

X-ray diffraction (with a CuK α radiation, Rigaku, D/Max-B) and scanning electron microscopy (FE-SEM Hitachi Su-70) techniques were used to identify the formed phases and inspect the film's microstructures. A detailed analysis of film / substrate interfaces was conducted by Rutherford Backscattering Spectroscopy (RBS). RBS experiments were carried out with 2 Schottky barrier detectors placed in IBM geometry at 140 ° and 180 ° scattering angles, with resolutions of 15 and 25 keV respectively, using a 2.0 MeV He^+ beam. Analyzes were performed with the samples tilted at 0 and 30 ° (the angle between beam direction and sample normal). All RBS data obtained for BSPT/Pt and BSPT/IrO_2 samples were simultaneously analyzed with the NDF program, 74 and the related depth profiles were derived to further study the microstructure and film / electrode interfaces. A modified commercial atomic force microscope (Multimode, Nanoscope SA, Veeco) was used for the surface roughness characterization.

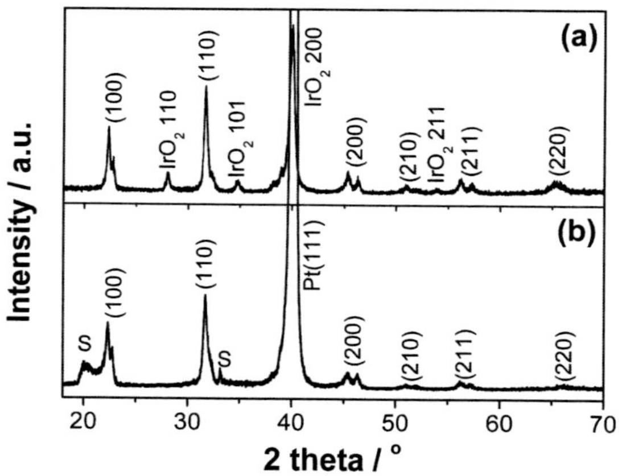

Figure 6.17. XRD patterns of sol-gel derived BSPT 37/63 films on $IrO_2/Ti/SiO_2/Si$ (a) and $Pt/Ti/SiO_2/Si$ (b). Peaks marked as "S" in the bottom curve denote the diffraction lines of $Pt/TiO_2/SiO_2/Si$ substrate. In the upper curve, 110, 101, 200 and 211 peaks of IrO_2 of the $IrO_2/TiO_2/SiO_2/Si$ substrate can be also noticed.

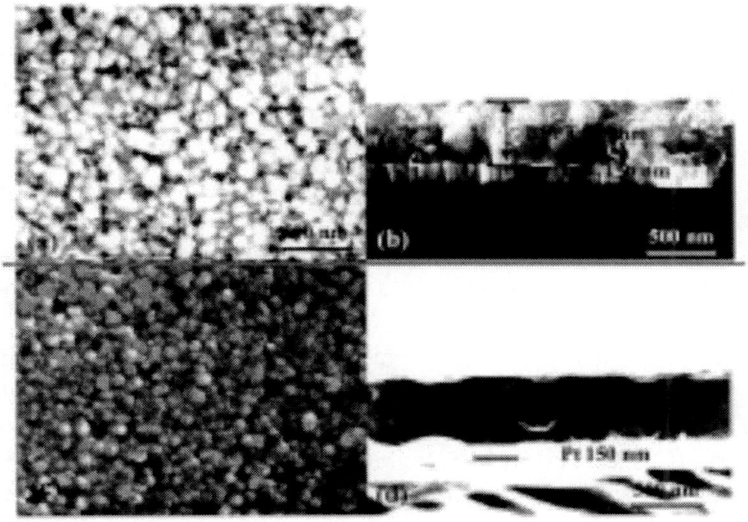

Figure 6.18. Top view ((a) and (c)) and cross-section ((b) and (d)) SEM micrographs of BSPT 37/63 films on $IrO_2/Ti/SiO_2/Si$ and $Pt/Ti/SiO_2/Si$. Grain sizes vary between 150 and 200 nm, and the total film thickness is about 550 nm.

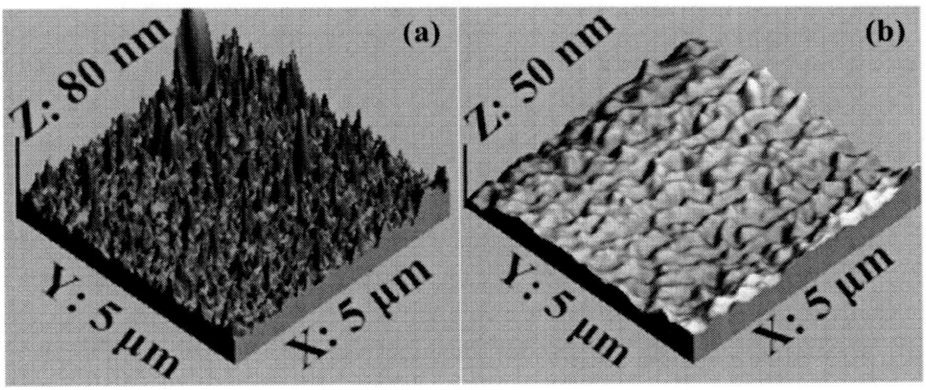

Figure 6.19. The AFM surface topographic images of BSPT/Pt (a) and $BSPT/IrO_2$ (b).

For electrical measurements, gold top electrodes with a diameter of 0.6 mm were deposited on the films by dc sputtering using a shadow mask followed by an annealing treatment at 300 °C for 30 minutes in air. Electrical measurements were performed on a number of BSPT/Pt and $BSPT/IrO_2$ samples, and on several capacitors per sample. The percentage of non short-circuited capacitors per sample is 15% higher for $BSPT/IrO_2$ films than for equivalent BSPT/Pt ones. The dielectric constant and loss of $BSPT/IrO_2$ and BSPT/Pt capacitors as a function of frequency and ac electric field was measured with an HP4284 Precision LCR Meter, from room temperature to 550 °C. The ferroelectric properties and current - electric field curves (I - V) were measured using a TF analyzer (TFA-1000) and a Keithley-617 electrometer at room temperature, respectively.

Figure 6.17 shows the XRD patterns of 550 nm thick BSPT/IrO_2 and BSPT/Pt films. Both exhibit pure perovskite phase with no evidence of any extra phases resulting from the decomposition or interaction between the film and the substrate. Based on the estimated values of full width at half maximum (FWHM) of (100) peaks, BSPT/IrO_2 films (FWHM: 0.50 °) show a good crystallinity comparable to BSPT/Pt films (FWHM: 0.54 °). No preferred orientation was observed for any of these films, apart from a different peak intensity for the 2-theta diffraction peaks splitting at around 45 ° and 55 °. The lattice constants of BSPT/Pt [a = 3.9695(5) Å, **c** = 3.9903(5) Å, c/a = 1.0052] and BSPT/IrO_2 [a = 3.9687(5) Å, c = 4.0050(5) Å, c/a = 1.0091] were obtained, which would seem to indicate a slightly higher degree of tetragonality for IrO_2 based films.

Figure 6.18 (a)-(d) presents SEM surface morphologies and cross-section microstructures of BSPT/IrO_2 and BSPT/Pt films. Both films are characterized by fairly dense micro-structures, with an average grain size of about 150 nm and a total thickness of about 550 nm. Figure 6.19 (a) and (b) show the AFM topographic images of BSPT/Pt and BSPT/IrO_2. The average surface roughness of these two films measured by AFM is about 4 nm over a 5 μm scan area. From the SEM cross-section images, the interface between BSPT films and the electrodes is clear, and no intermediate or inter-diffusion layer is observed.

RBS measurements were performed to further inspect the film / electrode interfaces. Figure 6.20 presents the simulated depth profiles of the different elements of BSPT/IrO_2 and BSPT/Pt films deduced from the RBS simulation analysis. According to the analysis, the RBS spectra can be accurately simulated using the normal composition for the BSPT layer, $0.37BiScO_3$-$0.63PbTiO_3$. According to the RBS results the total area densities of the BSPT layer and $PbTiO_3$ seed layer on IrO_2 and Pt electrodes are 25.3×10^{17} atoms cm^{-2} and 20.5×10^{17} atoms cm^{-2}, respectively.

In Figure 6.20, Si (1) denotes Si elements in the substrate and Si (2) denotes the elements in the SiO_2 buffer layer, Ti (1) denotes Ti elements in the TiO_2 buffer layer, Ti (2) denotes Ti in the BSPT films, and Ti (3) (in Figure 6.20 (a)) denotes Ti at the interface between the ferroelectric film and the IrO_2 electrode. Figure 6.20 (a) clearly indicates the existence of element mixing at the film / electrode interface in BSPT/IrO_2 films. A clear thin interface region between the IrO_2 electrode and the film is observed in the RBS spectra, that was not detected by SEM (Figure 6.18 (b)) and which results from the inter-diffusion reaction between the underlying IrO_2 electrode and the ferroelectric film. This region has a real density of approximately 2.0×10^{17} atoms cm^{-2}, corresponding to about 8% of the total ferroelectric film thickness. Considering the total thickness of the films measured by SEM, this reaction layer could be up to 44 nm in depth. This estimate assumes that the interface has the same density as the BSPT film although the real interface thickness may vary if the density of the interface differs from the BSPT film. On the other hand, Figure 6.20 (b) indicates that BSPT/Pt samples are characterized by sharp and flat interfaces, without any interface mixing between the top ferroelectric layer and Pt electrode.

The observed differences in the film / electrode interfaces of BSPT/IrO_2 and BSPT/Pt are indicated by the schematics shown in Figure 6.21, which may affect the electric performance of BSPT films.

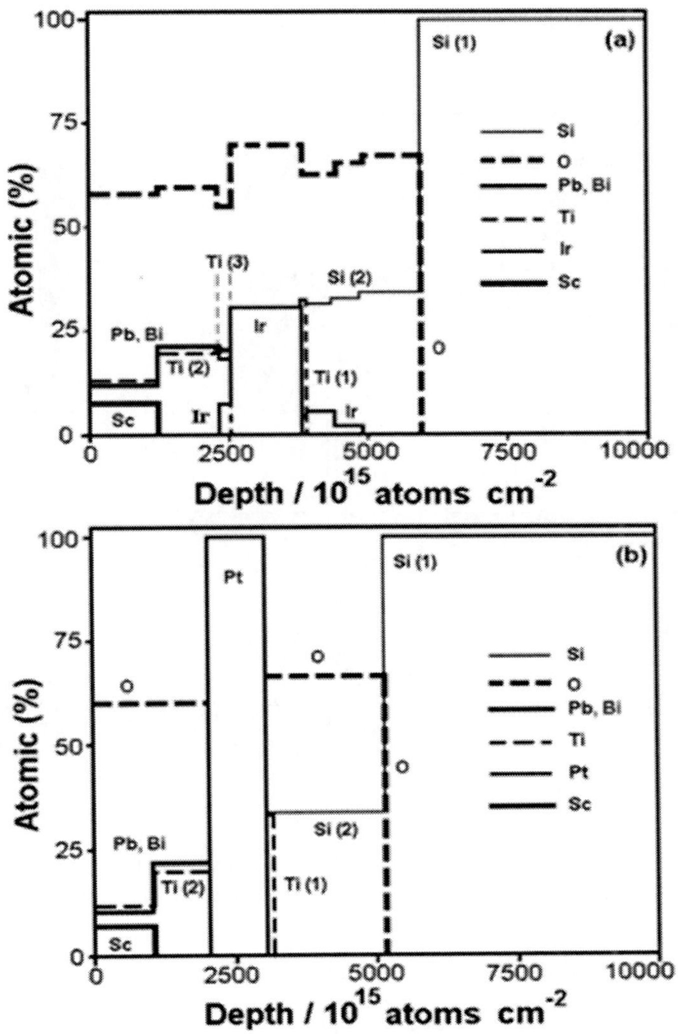

Figure 6.20. Depth profiles of the different elements of BSPT/IrO_2 (a) and BSPT/Pt (b) films deduced from the simulation analysis of RBS results. Si (1) denotes Si elements in the substrate and Si (2) denotes the elements in the SiO_2 buffer layer. Ti (1), (2), and (3) denotes respectively Ti elements in the TiO_2 buffer layer, the BSPT films, and the interface between the ferroelectric layer and the IrO_2 electrode.

The frequency dependence of the dielectric constant and dielectric loss at room temperature is depicted in Figure 6.22, from 100 Hz to 1 MHz. The average relative errors associated with the dielectric constant measurements of BSPT/IrO_2 and BSPT/Pt films are less than 0.025 and 0.0305, respectively as indicated by the error bars in the figure and implying a good reproducibility of the dielectric measurements for the studied films. At 100 Hz and at room temperature, BSPT/IrO_2 and BSPT/Pt films exhibit very similar dielectric properties, a high dielectric constant of about 1500 and 1537, and low dielectric loss <0.027 and 0.030, respectively, that deviates as the frequency increases. The figure clearly shows that

$BSPT/IrO_2$ films show a higher dependence of the dielectric response on the frequency compared to BSPT/Pt films.

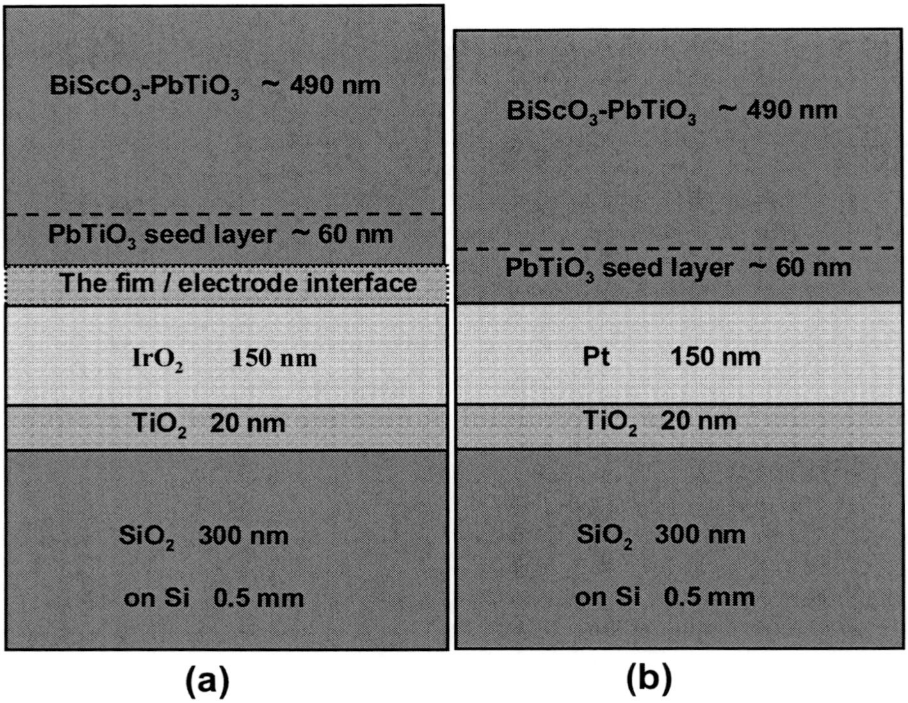

Figure 6.21. Schematic layer characteristics of $BSPT/IrO_2$ (a) and BSPT/Pt (b). In compared to BSPT/Pt, $BSPT/IrO_2$ exhibits an evident film/electrode interface, which leads to improvement in the electrical properties.

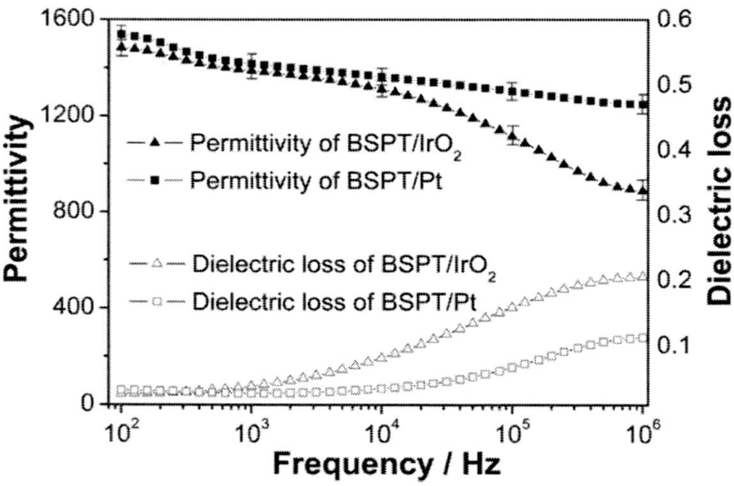

Figure 6.22. Frequency dependence of the permittivity and dielectric loss of $BSPT/IrO_2$ and BSPT/Pt films.

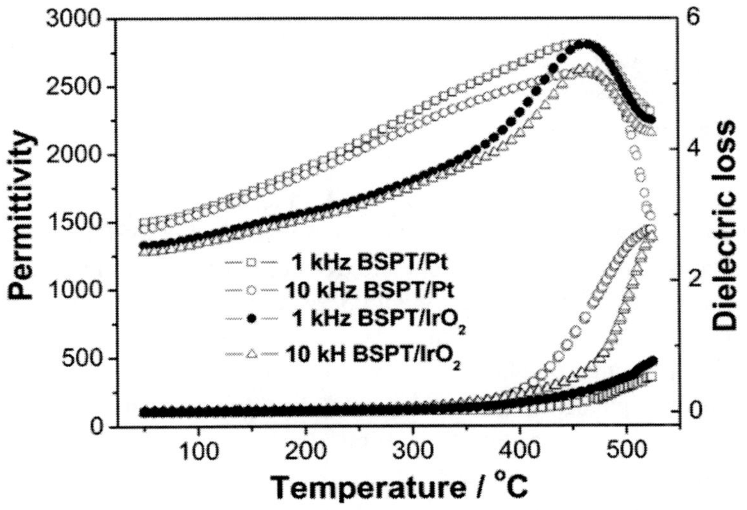

Figure 6.23. Temperature dependence of the dielectric permittivity and dielectric losses of BSPT/IrO_2 and BSPT/Pt films.

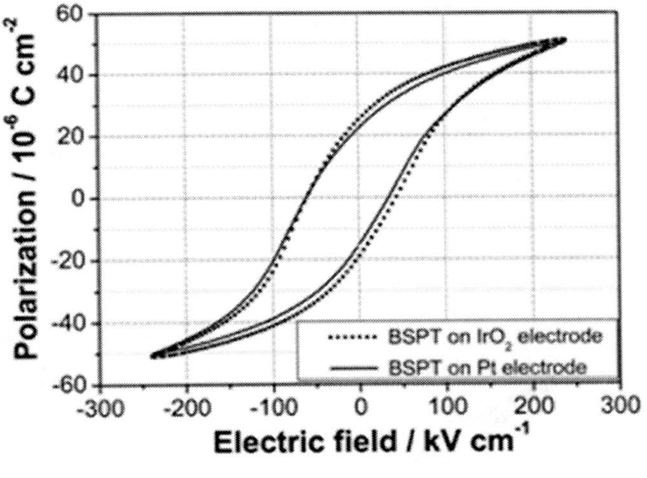

Figure 6.24. Polarization versus electric field of BSPT/IrO_2 (solid line) and BSPT/Pt (dashed line) films at 100Hz and at room temperature.

Figure 6.23 describes the temperature dependence of the dielectric properties of BSPT films on IrO_2 and Pt electrodes at 1 kHz and 10 kHz. The *Tc* of both films is around 460 °C, the temperature at which the dielectric constant exhibits a nearly identical maximum of 2670 and 2700 (at 1 kHz), for BSPT/IrO_2 and BSPT/Pt, respectively.

The measured *Tc* is consistent with reported values for bulk MPB BSPT ceramics, and is over 100 °C higher than that of MPB PZT thin films. It is also worthwhile noting that the dielectric losses of BSPT/Pt and BSPT/IrO_2 remain below 0.1 up to 470 °C and 400 °C, respectively, which is comparable to values previously reported for rhombohedral and tetragonal BSPT thin films, [75] and should facilitate the utilization of these films in high temperature applications. Nevertheless, a different curve profile for the dielectric permittivity response of BSPT/IrO_2 and BSPT/Pt films is clearly seen. For BSPT/IrO_2, the variation of the dielectric permittivity as a function of the temperature is similar to the one of equivalent BSPT ceramics, [76] with a marked increase of the dielectric permittivity near 400 °C. However, for BSPT/Pt films a more diffuse curve with a broad dielectric peak around 300 - 500 °C is observed, which is similar to the curves of BSPT films reported by H. Wen *et al*, Yoshimura *et al* and Nino *et al.* [75]

Well-defined hysteresis loops for BSPT/IrO_2 and BSPT/Pt were measured at room temperature and are shown in Figure 6.24. The hysteresis loop of BSPT/IrO_2 is slightly open along the polarization axis and exhibits to some extent a higher maximum polarization value, in comparison to that of BSPT/Pt. A remanent polarization (P*r*) of 26 μC cm^{-2} and a coercive field (E*c*) of about 40 kV cm^{-1} were calculated for BSPT/IrO_2 films, in comparison with values of about 23μC cm^{-2} and 33.6 kV cm^{-1} for BSPT/Pt. The IrO_2 electrode slightly increases the remanent polarization as well as the coercive field, but does not substantially affect the ferroelectric behavior of the BSPT films.

Figure 6.25 depicts the leakage current behavior, i.e. J_L – E curves (I – V curves) of BSPT/IrO_2 and BSPT/Pt films. Both J_L – E curves show a rectifying behavior, i.e., asymmetric leakage currents for a negative and positive electric field. [77] This asymmetry can be attributed to the presence in these films of different top and bottom electrodes and related to Scottky emission. [78, 79] The J_L – E curve of BSPT/IrO_2 exhibits two Ohmic regions (1) and (2) from 0 to 80 kV cm^{-1}, where the leakage current linearly increases with the applied electric field. On the other hand, in the J_L – E curve of BSPT/Pt films along the positive voltage direction, four regions can be identified: (1) the Ohmic region, where the conductivity is linear; (2) a nearly constant region; (3) a nonlinear region; and (4) a vertical current enhancement region. The obvious differences in the leakage current characteristics of BSPT films deposited on IrO_2 and Pt electrodes point to the differences in the conduction behavior between these films, suggesting differences in the space charge distribution at the film / electrode interfaces for both capacitors.

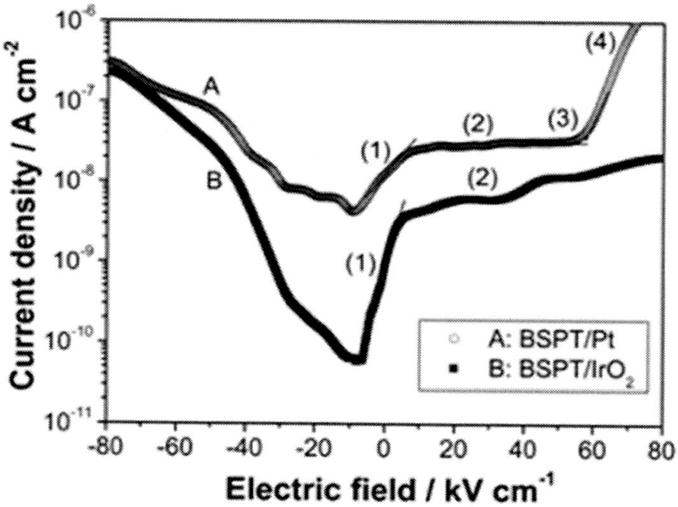

Figure 6.25. The relation between the current density and dc electric field or applied voltage of BSPT/Pt films (a) and BSPT/IrO_2 (b). The maximum applied voltage is 4 V.

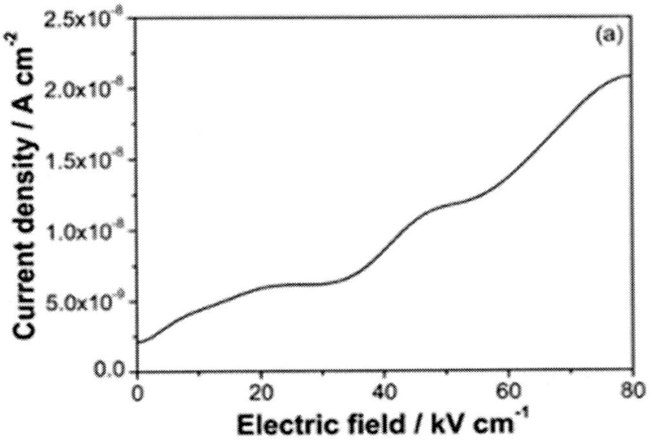

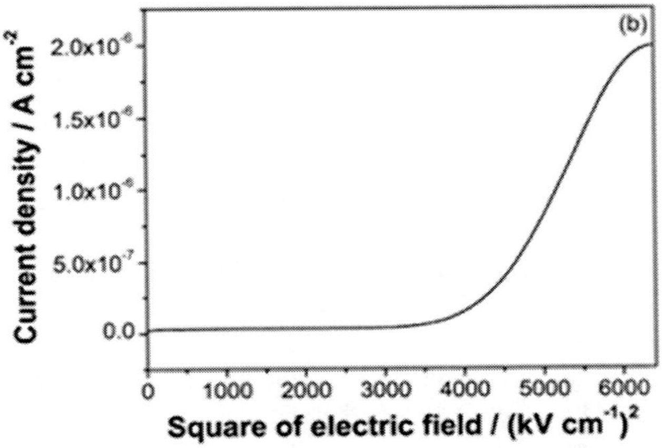

Figure 6.26. Current density (J-E) versus electric field of BSPT/IrO_2 thin films at positive voltages (a) and current density (J-E) versus the square of electric field (J-E^2) characteristics of BSPT/Pt thin films at positive voltages (b).

Particularly encouraging are the low leakage current values of BSPT/IrO_2 capacitors. In the low electric field region, up to +54 kV cm^{-1}, the leakage current of BSPT/Pt is 4 to 10 times higher than that of BSPT/IrO_2 (Figure 6.25). Above +54 kV cm^{-1}, the leakage current density of BSPT/Pt increases markedly with the applied electric field. In high electric field regions (+65 ~ +80 kV/cm), the leakage current density of BSPT/IrO_2 is of the order of $J_L \leq 10^{-8}$ A cm^{-2} for BSPT/IrO_2, whereas that of BSPT/Pt exceeds 10^{-6} A cm^{-2}. These results clearly imply a better leakage current performance of BSPT/IrO_2 capacitors over BSPT/Pt ones. Additionally, the increase of the leakage current density of BSPT/IrO_2 with the electric field shows a good linearity and no drastic breakdown occurs, within the studied field range.

From the above results, the use of IrO_2 electrodes does not significantly affect the ferroelectric properties of BSPT capacitors, but improves the temperature dependence of the dielectric permittivity and the leakage current performance, demonstrating the positive effect of the film / electrode interface, that as a consequence increases the electrical reliability of BSPT films on IrO_2.

It is well known that the electric properties of ferroelectric films depend on many factors, which include the degree of crystallization, crystal structure and film thickness among others. Because all these aspects are similar in BSPT/IrO_2 and BSPT/Pt films, the dissimilarities observed in the electrical response, in particular the leakage behavior should be ascribed to the different contributions of the bottom electrodes, i.e. to the interface formed with the ferroelectric film. The observed high leakage current density of BSPT/Pt capacitors points to the existence of a space charge contribution possibly related to the film / electrode interface when Pt is used as the bottom electrode.

A closer analysis of the leakage current behavior for both capacitors is depicted in Figure 6.26 (magnification of Figure 9 for the positive voltage side). As shown in Figure 6.26 (a), BSPT/IrO_2 exhibits a nearly linear J_L - E relationship, described as:

$$J_L = \frac{n_0 q \mu E}{S} \tag{1}$$

where J_L, n_0, q, μ, E, and S stand for the leakage current density, refractive index, charge of electrons, drift mobility of charge carriers, applied electrical field, and area of top electrode, respectively, which indicates that the leakage current behavior in BSPT/IrO_2 is described by an Ohmic conduction mechanism and the contribution of space charges for the conduction process can be neglected. [80]

However, a different situation was observed for BSPT/Pt capacitors. Figure 6.26 (b) describes the $J_L - E^2$ relation for BSPT/Pt films in which two conduction regimes are clearly observed. For low voltages, the linear current – voltage variation of BSPT/Pt is also described by an Ohmic type conduction mechanism. When the voltage is higher than 3.2 V (V2>10V), a $J_L - E^2$ relationship holds, which indicates that the leakage current of BSPT/Pt is ruled by a mechanism of space-charge limited currents (SCLC) induced by the existence of space charges, [78] and described by the formula:

$$J_L = \frac{9 \mu \varepsilon_0 \varepsilon_r E^2}{8d} \tag{2}$$

where ε_0 stands for the permittivity of free space, ε_r for the permittivity of the dielectric, d for the thickness of the film. As the voltage increases, the injected charges increase and J_L follows a E^2 dependence rather than an Ohmic relationship. Thus, the contribution of space charges for the conduction process of BSPT/Pt films cannot be neglected.

It is also possible to estimate the drift mobility (μ) of charge carriers as a function of the applied voltage for both capacitors. According to the formulas (1) and (2), the μ value of BSPT/IrO_2 is estimated to be of the order of 10^{-8} cm^2 V^{-1} s^{-1}, while that of BSPT/Pt to be of the order of 10^{-12} cm^2 V^{-1} s^{-1}. These different values indicate that for positive voltages, the charge carriers in BSPT/IrO_2 move quickly at a much higher rate than that of the carriers in BSPT/Pt films, which suggests that the dominant charge-carrier species in BSPT/IrO_2 are free electrons, while for BSPT/Pt they are possibly associated with lattice defects such as oxygen vacancies. A charge mobility of 2×10^{-12} cm^2 V^{-1} s^{-1} was also estimated for $Ba_{0.5}Sr_{0.5}TiO_3$ films and associated with oxygen vacancies. [81]

The probable presence of space charges at the film/electrode interface of BSPT/Pt films is also reflected in the dielectric permittivity characteristics. As shown in Figure 6.23, the profile of the temperature dependence of the permittivity of BSPT/IrO_2 is more similar to the equivalent BSPT ceramics, whereas in comparison that of BSPT/Pt shows a broad profile, which is possibly ascribed to the contribution of space charges. As the temperature increases, the dielectric permittivity of BSPT/Pt film also deviates from that of BSPT/IrO_2 thin film and it becomes similar only at the phase transition temperature, the temperature at which the polarization is maximized and its effect overcomes the space charge. Meanwhile, the dielectric loss values of BSPT/IrO_2 and BSPT/Pt films are almost the same at room temperature.

The space charge effect in BSPT/Pt may originate from its film/electrode interface, which is quite different from that of BSPT/IrO_2. Indeed, according to the microstructure analyzes by SEM and RBS, BSPT/Pt shows sharp and flat interfaces, without any interface mixing between the top film and Pt electrode, while for BSPT/IrO_2 a mixed film / electrode interface was identified by RBS. In this case, a thin intermediate layer is assumed to be formed by the inter-diffusion between the underlying IrO_2 electrode and the BSPT thin film elements due to the oxide nature of both and with an average value of ~ 8% of the total film's thickness.

Space charge layers have been reported near the surface of ferroelectric films, [82] and arise from the trapping of electrons at surface traps or vacancies. In the present case of BSPT films, it is very possible that intrinsic defects such as oxygen vacancies trap electrons, and generate space charges, as well.

Under an electric field, the charged particles will be oriented and tend to accumulate near the film / electrode interface. The presence of a thin intermediate layer formed at the interface may act as a place to null the space charges; the following possible reaction may take place at the interface during the thermal treating process or under an electric field for the case of BSPT / IrO_2 capacitors: [83]

$$MO \rightarrow MO_{1-y} + \frac{1}{2} y O_2 \uparrow \quad (3)$$

i.e.,

$$IrO_2 \rightarrow IrO_{2-y} + \frac{1}{2} y O_2 \uparrow \quad (4)$$

where M stands for the cation and O the anion, Ir stands for the iridium cation, IrO_{2-y} corresponds to a vacant oxygen lattice site. Formula (4) may be rewritten as

$$O_{IrO_2} \rightarrow V_{IrO}^{2-} + e + O \quad (5)$$

where the O ions generated in the intermediate layer may neutralize the ionized oxygen vacancies nearest the film / electrode interface, whereas, the oxygen deficiency of the newly generated V_{IrO}^{2-} vacancies may be compensated by the oxygen atoms migrating from the IrO_2 electrode. Therefore, the IrO_2 electrode can reduce or eliminate the accumulation of space charges near the film / electrode interface, displaying a "sink defect effect".

In comparison, Pt electrodes have a much poorer affinity with oxygen and possess a very low tolerance limit for oxygen vacancies. [66] In addition, no marked intermediate layer was found near the film / electrode interface indicating that reactions similar to (4) and (5) will not take place. Pt electrodes will not promote a "sink defect effect", and space charges will accumulate near the film / electrode interface of BSPT/Pt films.

It is also proposed that the existing space charges contribute to the slight increase of the dielectric permittivity of BSPT/Pt films, which will become even more active at high temperatures, as observed in Figure 6.22. Thus a "sink defect model" is suggested to explain the difference in the conduction behavior of BSPT/Pt and BSPT/IrO_2 films.

As mentioned the formation of an intermediate layer between the film and IrO_2 electrodes was reported for PZT capacitors as well [83]; ferroelectric properties significantly improved and the dielectric properties were reported as inferior, when compared with films on Pt. However, in the present study, IrO_2 electrodes do not substantially affect the ferroelectric response of BSPT films. As shown in Figure 6.23, the use of IrO_2 slightly increases both the remanent polarization and the coercive field of BSPT films in comparison with BSPT/Pt. This phenomenon can be also attributed to the accumulation of space charges and oxygen vacancies near the film / electrode interface in BSPT/Pt that may lead to a suppression of the degree of switched polarization, and an increase in coercive fields as observed before. [84, 85] Another significant difference from previously reported cases is that in the present study the intermediate layer has a marked positive effect on the leakage current behavior and dielectric response of the capacitors fabricated on oxide electrodes.

It is therefore expected that the "sink effect of space charges" will lead to an improvement of the ferroelectric fatigue performance of BSPT/IrO_2 thin films as well. Fatigue studies are now being conducted.

In a simple summary, the effect of IrO_2 electrodes on the electric performance of sol-gel derived BSPT (37/63) thin films was investigated. Similar dielectric and ferroelectric behavior was observed for BSPT capacitors on IrO_2 and Pt electrodes. However a marked reduction in the leakage current density from 10^{-6} A cm^{-2} for films on Pt to 10^{-8} A cm^{-2} for films on IrO_2 occurred. RBS analysis indicate an interface reaction with the formation of an interface layer between the ferroelectric film and IrO_2, whereas BSPT films on Pt present a sharp and flat film / electrode interface. It was then proposed that this interlayer resultant from the reaction between the oxides of the film and the oxide electrode acts as a "sink" for space charges existing in the film, contributing to the improvement of the reliability of these capacitors.

6.4. LNO Electroded $0.37BiScO_3$-$0.63PbTiO_3$ Ferroelectric Thin Films

This section addresses the structure and electrical properties in sol-gel derived morphotropic phase boundary (MPB) $BiScO_3$-$PbTiO_3$ (BSPT) thin films, through a strategy of film quality tailoring by introducing a $LaNiO_3$ (LNO) multifunctional layer between the ferroelectric films and Pt / TiO_2 / SiO_2 / Si substrates. We demonstrated that this strategy has a desired positive effect on the film quality improvement and property enhancement for the BSPT films, indicating that the films on LNO layers exhibit a highly (100)-oriented texture and more excellent electrical properties than those on the pure platinized silicon substrates. This effect is ascribed to the integration in LNO layers with the functions of seed layers, oxide electrodes, stress buffer layers, and textured templates, which can overcome the disadvantages that appeared in the sol-gel fabrication of BSPT thin films on platinized silicon substrates.

To meet the miniaturization requirements of the microelectronics industry, further developments on BSPT films are also desired for being incorporated into IC circuitry. Although this topic is very popular, the miniaturization requirements with greater functionality in microelectronics applications continue to drive the development of better BSPT thin films with enhanced dielectrical and ferroelectric properties. For achieving this goal, the problems lying beneath the film fabrication and severely restricting their potential utilization have to be solved.

Generally, ferroelectric thin films possess poorer dielectric and ferro/piezoelectric properties in comparison with the ferroelectric ceramics and single crystals, [86] which is commonly ascribed to the sensitive nature of film quality and structure to many parameters other than composition and film thickness, such as film defects, substrate effects (residual stress), electrode effects, and crystallographic orientation. [87] These parameters contribute to many problems for controlling the film structure and the properties in fabrication processes. In general, these problems can be strategically dealt with, by introducing a functional interlayer between the film and Pt electrode substrate.

The first possible strategy is to use a seed layer to control the nucleation process, promote the phase formation, improve the film crystalline degree, depress the film defects, and prevent the reaction between films and substrates. [88-91] In this way, one can solve the problems of micro-structural heterogeneities and variations in crystalline quality that lead to interference with the desired ferroelectric properties. [92] On the other hand, many ferroelectric thin films deposited on $Pt/Ti/SiO_2/Si$ often exhibited bad interfacial effects or undesired inter-diffusion processes at the interface during the fabrication process, contributing to the poor electrical properties and serious fatigue or other reliability problems. As a result, the second possible approach is to introduce an oxide electrode layer replacing the metal one to improve the performance. [93-96] The third strategy involves choosing appropriate inter-layers to serve as buffer layers between the ferroelectric film and platinized silicon substrate to effectively modify the residual stresses [97] which also have a significant effect on the ferro / piezoelectric and dielectric film properties. Finally, the most key pathway is to use an oriented texture template inserted between the platinized silicon substrate and the ferroelectric film to control the film orientation and thoroughly optimize the electrical properties. [98, 99]

These distinct solutions are supposed to address the issues of poor crystalline and poor electrical properties in BSPT thin films. The more attractive commercially and clever alternative way is using an oxide multi-function-layer that can integrate the different effects of the seed layer, oxide electrode layer, buffer layer, and texture template layer, to improve the dielectric and ferro / piezoelectric properties of BSPT thin films on platinized silicon substrates. However it is very challenging to find or select a proper oxide material to serve such a multi-function-layer for this fulfillment, as a novel conductive oxide, $LaNiO_3$ (LNO) seems to be a good strategic choice. [100-102] It exhibits low and isotropic resistivity, and can act as a source of oxygen during heat treatment that prevents the inter-diffusion of vacancies between the ferroelectric film and substrate, and facilitates the nucleation of BSPT at its surface, which makes it an excellent oxide bottom electrode and seed layer for the growth of BSPT thin films. LNO possesses a pseudocubic peroveskite structure (a ~ 0.384 nm), that is lattice matching and structurally compatible with BSPT thin film (a ~ 0.405 nm) and $Pt/TiO_2/SiO_2/Si$ substrate (lattice constant of the out-of-plane is about 0.39 nm) together. This also allows us to tailor the residual stress, control and enhance the film orientation of BSPT thin films. Previously, LNO had been adopted to achieve highly textured PZT thin films. [103] In the other case of (Pb, La)TiO_3 (PLT) thin films, LNO layers were introduced between the ferroelectric films and the oxidized silicon substrates, for serving three functions as bottom electrodes, seeding layers, and composition diffusion barrier, to enhance the electrical properties. [104]

Therefore in this work, LNO layers will be fabricated on the $Pt/TiO_2/SiO_2/Si$ substrate by a sol-gel technique for serving as an interlayer between the ferroelectric film and Pt electrode, for the further sol-gel deposition of BSPT (37/63) thin films. The effects of LNO layers on the structural and electrical performance of sol-gel derived BSPT thin films were explored. In this strategy, LNO serves multi-functionality as oxide bottom electrode, seed layer, buffer layer, and especially an epitaxial template, which is expected to significantly enhance the material's electrical properties, and open a new vista in the field of growth and application of BSPT ferroelectric films or other functional systems.

Materials. Lanthanum nitrate hexahydrate [$La(NO_3)_3 \cdot 6H_2O$] (Merck, 99%) , nickel (II) nitrate hexahydrate [$Ni(NO_3)_2 \cdot 6H_2O$] (Fluka, 98.5%), ethanol absolute (Merck, 99.8%) and formamide (Merck, 99.5%), lead acetate ($Pb(COOCH_3)_2$) (Panreac Company, 99.5%), bismuth nitride ($Bi(NO_3)_3$) (Sigma-Aldrich, 99.99+%), scandium acetate ($Sc(COOCH_3)_3$ (Sigma-Aldrich, 99.9%), titanium isopropoxide (Ti(i-Pr)$_4$) (Sigma-Aldrich, >97%), glacial acetic acid (Pronalab, 100%), 2-propanol solutions (Fluka, 99.8%), acetylacetone (Merck, 99.5%), deionized water, $Pt/TiO_2/SiO_2/Si$ platinized silicon substrates.

Deposition of $LaNiO_3$ Films. Based on a wet chemical solution route, [105] the deposition of conductive LNO films on $Pt/TiO_2/SiO_2/Si$ was carried out. Lanthanum nitrate hexahydrate and nickel (II) nitrate hexahydrate were dissolved respectively in deionized water, and mixed in the stoichiometric ratio of La:Ni=1:1. In order to improve the solution wettability to the substrate and to avoid the film cracking during heating, ethanol absolute and formamide were also added, till the resultant concentration of LNO solution was diluted to 0.2 M. The LNO films were deposited by spinning the solution on $Pt/TiO_2/SiO_2/Si$ substrates at 4500 rpm for 40 seconds and annealed at 700 °C for 30 minutes.

Deposition of $BiScO_3$-$PbTiO_3$ (BSPT) Films. A 0.2M precursor solution with the composition ratio of $0.37BiScO_3$-$0.63PbTiO_3$ [BSPT (37/63)] was synthesized by dissolving lead acetate, bismuth nitride, scandium acetate, and titanium isopropoxide into the solvents composed of glacial acetic acid, 2-propanol solutions, deionized water, and acetylacetone. Formamide was used as an additive to improve the adhesion of the film to the substrate. The as-grown $LaNiO_3/Pt/TiO_2/SiO_2/Si$ (200nm/150nm/20nm/300nm/Si substrate) was used as substrate. BSPT films were spin coated with a spin rate of 3000-5000 rpm and a spin time of 30 seconds. Each layer was dried at 350 °C and pyrolyzed at 650 °C for 5 minutes. After all the 10 layers had been deposited, the BSPT films on $LaNiO_3$ electrodes (hereafter abbreviated as BSPT/LNO) with the desired thickness (~550 nm) were heat treated at 750 °C for 2 hours. For comparison, BSPT thin films on $Pt/TiO_2/SiO_2/Si$ (150nm/20nm/300nm/Si substrate) (hereafter abbreviated as BSPT/Pt) were prepared under identical conditions as well.

Structural and Morphology Characterization. The structure and phase evolution of the films were analyzed by x-ray θ-2θ diffraction (XRD, with a CuK α radiation, Rigaku, D/Max-

B). X-ray rocking curves were obtained by tilting the BST films through the Bragg angle of the (200) plane. The X-ray pole figure measurements were performed with a Philips XPert MRD diffractometer using a Cu KR X-ray source. In the pole figure measurements, the diffracted intensity was collected at a step of 5° in the tilting and rotating angles in the whole hemisphere, at a fixed θ-2θ angle of the open detector, that corresponds to the (200) reflections of BST films on Pt and LNO, while the specimens were tilted and azimuthally rotated in relation to the incident beam. The X-ray rocking curves and pole figures were plotted to further highlight the film texture degree. The surface and cross-sectional microstructure of the films were observed by scanning electron microscopy (SEM).

Dielectric and Ferroelectric Measurements. For electrical measurements, gold top electrodes with a diameter of 0.6 mm were deposited on the films by dc sputtering using a shadow mask followed by an annealing treatment at 300 ℃ for 30 minutes in air. The dielectric constant and loss of BSPT/LNO and BSPT/Pt capacitors as a function of frequency at room temperature was measured with an HP4284 Precision LCR Meter. The ferroelectric properties were measured using a TF analyzer (TFA-1000) at room temperature.

Local Piezoelectric Measurements. A modified commercial atomic force microscope (Multimode, Nanoscope SA, Digital Instruments) was used in the experiments for investigating the local piezoelectric response. A conductive Pt coated *Si* tip–cantilever (EFM-20, Resonance frequency: 75 kHz / Force constant: 2.8 N/m) system was used for the application of external voltages and for vibration detection. Ferroelectric thin films were excited by an external ac voltage V_{ac} applied between the PFM tip and the bottom electrode, and the deflection signal from the cantilever was detected by a lock-in amplifier. A topographic image of the film surface was taken simultaneously with the domain image. In the piezoelectric image, domains with opposite polarities exhibit different contrast.

The local piezoelectric hysteresis loops (piezoloops) were measured inside individual grains. The loop acquisition consists of applying a *dc* voltage, V_{dc}, for a short time and measuring the effective piezoelectric response between the pulses. In this work, the absolute values of d_{33} were not determined, and all data are given for the relative piezo-response values measured in volts (output analogue signal of the lock-in amplifier). Since the results were obtained with the same cantilever and identical scanning and acquisition conditions, the comparison could be made of the relative values of local piezoelectric coefficients for BSPT films on Pt and LNO electrodes.

Through choosing a sol-gel derived $LaNiO_3$ (LNO) layer as a multi-function-layer to implement the film tailoring strategy, morphotropic phase boundary (MPB) $BiScO_3$-$PbTiO_3$ (BSPT) thin films had been fabricated by chemical solution deposition methods. The influence of this layer on the film crystallization, texture control, and electrical properties was investigated, as described in the following sections.

XRD Analysis. Figure 6.27 shows the XRD patterns in θ-2θ scans of BSPT films grown on Pt and LNO layers, the full width at half maximum (FWHM) of (100) diffraction peaks were respectively 0.30^{o} and 0.24^{o}. These imply the introduction of LNO conductive layer between BSPT film and Pt electrode significantly improves the crystalline degree. On the other hand, in comparison with a random BSPT / Pt film, a highly (h00) textured BSPT / LNO film is achieved. A Lotgering factor (*f*) as following can be used to estimate the (100)-textured degree: [106]

$$f = \frac{p - p_0}{1 - p_0},$$

$$p = \frac{\sum I(h00)}{\sum I(hkl)},$$

where $p = p_0$ is the orientation intensity ratio for a randomly oriented sample. The f factors of BSPT/Pt and BSPT/LNO thin films were calculated to be 0.62 and 0.94, respectively, which indicates that BSPT/LNO thin films possess a much higher (h00)-oriented degree than BSPT thin films directly deposited on Pt electrodes. In addition, according to Figure 1, the observed interface relationship in BSPT/LNO is (100) BSPT // (100) LNO // (111) Pt, suggesting that the ferroelectric films have an epitaxy nature with respect to the LNO interlayer, which underlies beneath the excellent lattice matching occurred at the interface between perovskite BSPT (a ~ 0.405 nm) and LNO (a ~ 0.384 nm) films. Therefore, the LNO layer acts not only as an oxide electrode and seed layer, but also as an orientation template, enhancing the (h00) orientation degree for BSPT/LNO films.

X-ray Rocking Curves and Pole Figure Characterization. To better understand the preferred (h00) orientation of BSPT /LNO films, XRD rocking curves and pole figure measurements had been conducted respectively. As Figure 6.28 shows, in the rocking curves, BSPT/LNO films demonstrated an intense and narrow (200) peak in contrast to BSPT/Pt, indicating a high degree of preferred (h00) orientation of BSPT/LNO film. Figure 6.29 depicted the X-ray pole figure distribution of BSPT/LNO and BSPT/Pt, respectively corresponding to (200) reflections. In BSPT/LNO, only a strong diffraction point appears at the center position of (200) ring-type reflection peak, and shows a strong film texture. While in the case of BSPT/Pt, distinct diffraction concentrations not only appear at the peak center, but also spread over the ring bands, which shows a weak texture in (200) diffractions. These results are consistent with what is shown in Figure 6.27, confirming the (h00) preferred orientation of BSPT/LNO.

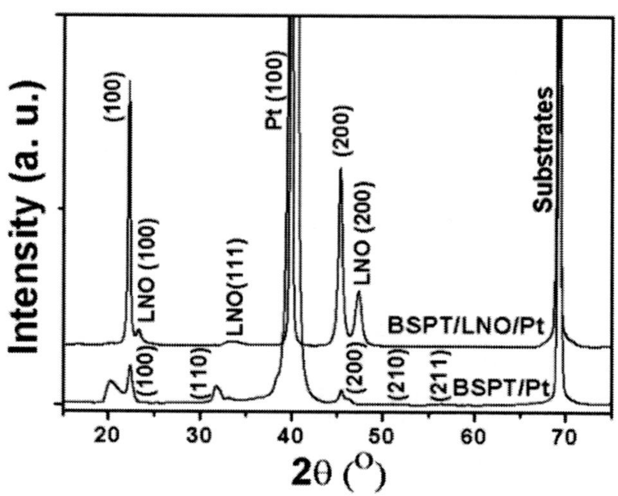

Figure 6.27. XRD curves of BSPT films grown on Pt and LNO layers.

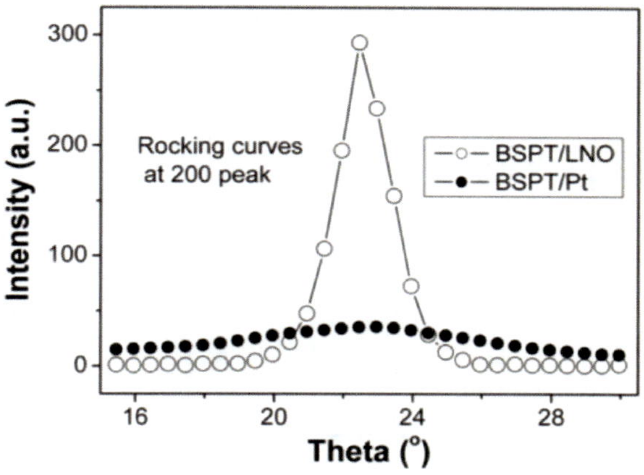

Figure 6.28. Rocking curves of BSPT thin films on Pt and LNO electrodes at (200) reflection.

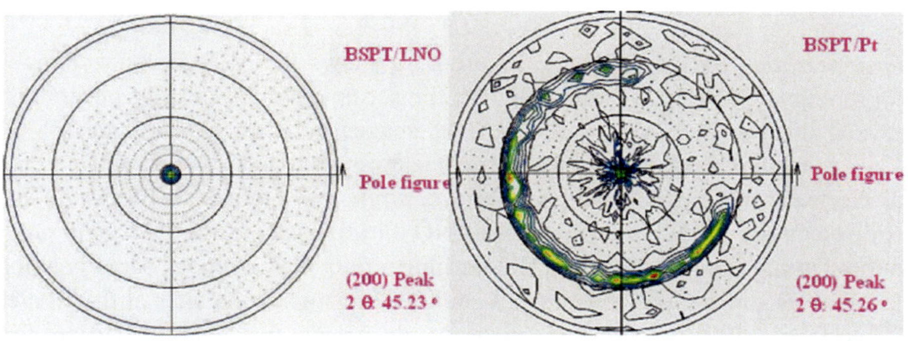

(a) (b)

Figure 6.29. X-ray Pole figures of BSPT thin films. (a)corresponding to (200) reflection (2θ=45.23°) of BSPT/LNO; (b)corresponding to (200) reflection (2θ=45.26°) of BSPT/Pt.

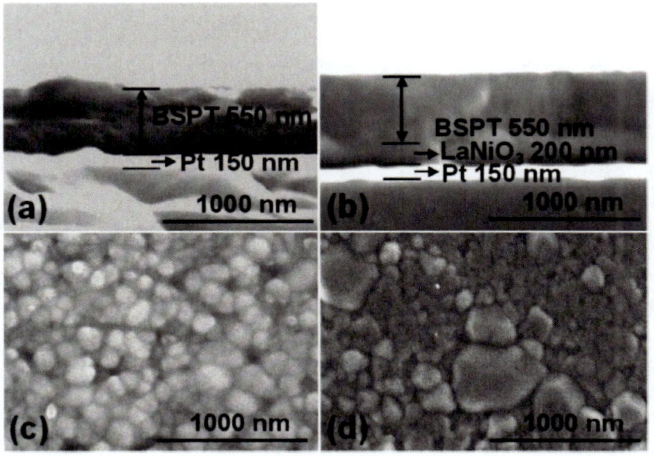

Figure 6.30. Cross-section ((a) and (b)) and top view ((c) and (d)) SEM micrographs of BSPT 37/63 films on $Pt/Ti/SiO_2/Si$ and $LNO/Pt/Ti/SiO_2/Si$. The total film thickness is about 550 nm.

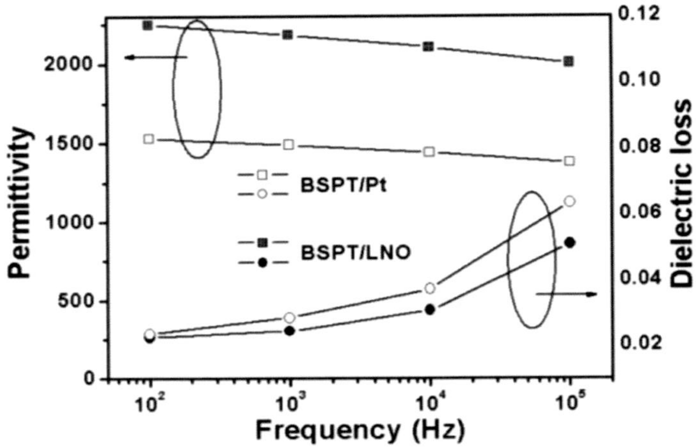

Figure 6.31. Frequency dependence of dielectric constant and dielectric losses of MPB BSPT (37/63) thin films on Pt and LNO electrode.

Characterization for Morphology. Figure 6.30 shows the SEM cross-sectional and top view morphology of BSPT/LNO and BSPT/Pt films. Some pinholes can be observed from the surface SEM images; the cross-sectional SEM images indicate that both the BSPT/Pt and BSPT/LNO thin films are fairly dense, with a thickness of about 550 nm. Between BSPT and Pt metal electrode there is an obvious interface, while the interface between the BSPT and LNO oxide electrode is blurry. The fact that LNO thin films act as the seed layer provides the explanation. During the growth of BSPT thin films, LNO also provides some positions (such as pinholes) at its surface and constitutes a multitude of nucleation sites at the film/electrode (LNO) interface, [107] and heterogeneously decreases the energy barrier for BSPT crystallization. This promotes the crystallization of ferroelectrics at the interface between BSPT and LNO, and makes the SEM observed interface rather vague. Whereas, the Pt cannot offer such type of positions at its surface, the crystal nucleus of BSPT is formed just at the inner of BSPT layer, so the interface between BSPT and Pt is sharply apparent.

LNO plays such an role of seed layer in achieving a high degree of crystallinity of BSPT/LNO films, which is in accordance with the decrease of the FWHM values of XRD peaks. In addition, from the cross-sectional images, the deposition of BSPT/LNO thin films is dominated by columnar growth rules, also leading to a higher orientation degree in BSPT/LNO rather than BSPT/Pt thin films. It is indicated that the LNO interlayer improves the epitaxial quality of BSPT thin films, which will enhance the dielectric, ferro/piezo-electric properties of the films.

Dielectric Measurements. Figure 6.31 depicts the frequency dependence of dielectric constant and dielectric losses of MPB BSPT (37/63) thin films on Pt and LNO electrodes. As expected, the BSPT films on LNO electrodes exhibit higher dielectric permittivity and lower dielectric losses than those on Pt electrodes. At room temperature, the BSPT/LNO films exhibit a permittivity as high as 2500 and a low dielectric loss of 0.02 at 100 Hz. This improved dielectric performance is also attributed to the effect from LNO as a seed layer and texture tailoring template, which improves the crystallinity and orientation degree associated dielectric properties of BSPT thin films.

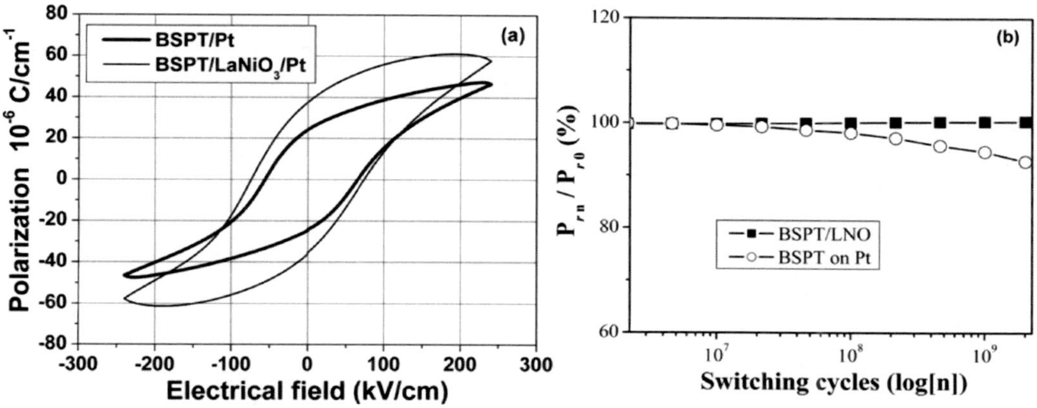

Figure 6.32. (a) Polarization versus electric field of BSPT/LNO and BSPT/Pt films at 100Hz and at room temperature; (b) Fatigue behavior of BSPT/LNO and BSPT/Pt till 2×10^9 switching cycles.

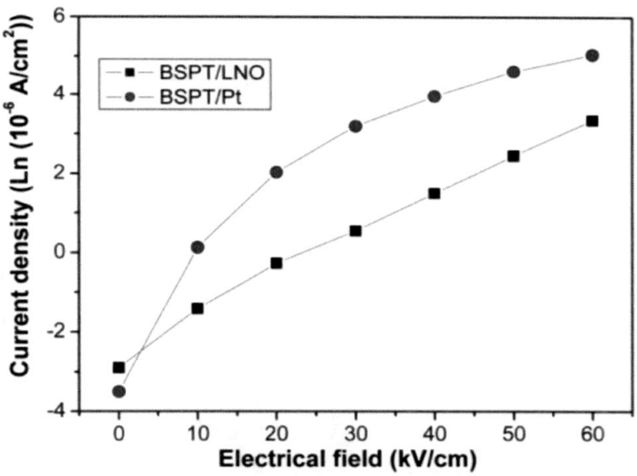

Figure 6.33. The leakage current density under different electrical fields of BSPT/LNO and BSPT/Pt thin films.

Polarization Measurements. Figure 6.32(a) shows the well-defined polarization (P-E) hysteresis loops for BSPT/LNO and BSPT/Pt measured at room temperature, under 100 Hz and a DC voltage of 12 V. A remanent polarization (Pr) of 40 μC cm^{-2} and a coercive field (Ec) of about 68 kV cm^{-1} were derived for BSPT/LNO thin films, in comparison with values of about 25μC cm^{-2} and 60 kV cm^{-1}, respectively for BSPT/Pt. The two loops show little shape difference. Although the introduction of the LNO electrode slightly increases the coercive field, it substantially enhances the remanent polarization of BSPT thin films and makes BSPT thin film show better saturated hysteresis loops, which indicates that this oxide electrode plays a positive effect on the ferroelectric properties.

This enhancement may be due to the fact that the LNO interlayer could act as a buffer and layer and an orientation template between BSPT film and Pt electrode, [15, 33-38] which can effectively modify the film strain. Generally, when the lattice parameter of buffer layers or substrates is smaller than that of films, a compressive strain will be induced in the thin films, which tends to elongate the lattice in the out-of plane direction, leading to the increase of remnant polarization. Based on the lattice parameters of Pt (~ 0.3900 nm), LNO (~ 0.384 nm), and BSPT (~ 0.405 nm), BSPT/LNO thin film is expected to exhibit a more compressive strain and then exhibits an enhanced remanent polarization, in comparison with BSPT/Pt thin films. On the other hand, in randomly orientated BSPT film on Pt metal electrode, the polarizations may cancel each other due to a different domain orientation direction. In highly orientated BSPT films on LNO oxide buffer films, the polarization value will increase due to a similar domain orientation direction.

Fatigue Behaviors. Figure 6.32(b) shows the fatigue behaviors of BSPT films on Pt and LNO electrodes, from which the measurements lasted up to 2×10^9 bipolar switching cycles. The ferroelectric films on Pt begin to exhibit fatigue behavior from 10^7 cycles, and after 2×10^9 cycles, apparent degradation of remnant polarization occurred. In comparison, no decrease of remnant polarization was observed for the BSPT/LNO films until 2×10^9 cycles. This indicates that, as an oxide electrode replacing Pt, the LNO layer significantly improves the reliability of BSPT films. The mechanism for the reliability improvement of BSPT films by LNO oxide electrode will be published elsewhere.

Leakage Current. The leakage current density under different electrical fields of BSPT/LNO and BSPT/Pt thin films is shown in Figure 6.33, which indicates that the introduction of the LNO electrode can improve the leakage current behavior and decrease the value of the current density.

Local Piezoelectric Measurements. To study the effect of the LNO electrode on enhancing the piezoelectric performance of BSPT thin films in light of the microscopic viewpoint, we investigated the local piezoelectric response by the method of piezo-force response microscopy (PFM). Figure 6.34 (a) and (c) show the topography images of BSPT/Pt and BSPT/LNO films, which indicate an average grain size less than 100 nm in BSPT/LNO films.

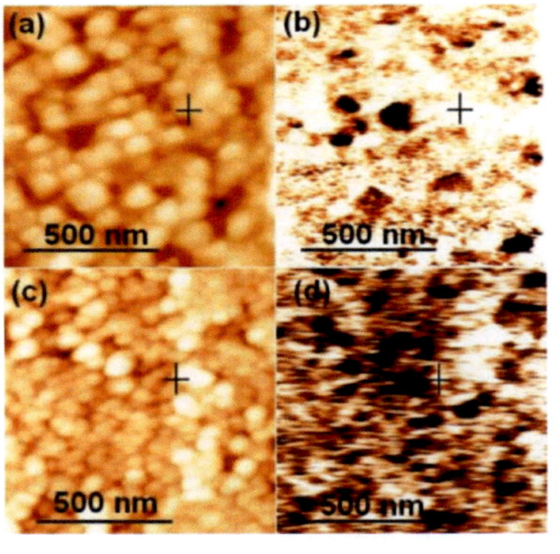

Figure 6.34. Topography and domain structures of BSPT/Pt and BSPT/LNO obtained by the Piezo Force Microscopy (PFM) technique.

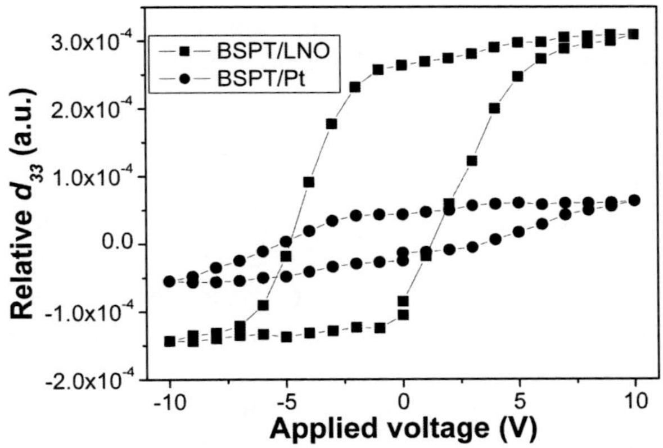

Figure 6.35. Local piezoelectric hysteresis loops (piezoloops) of BSPT/Pt and BSPT/LNO measured by PFM.

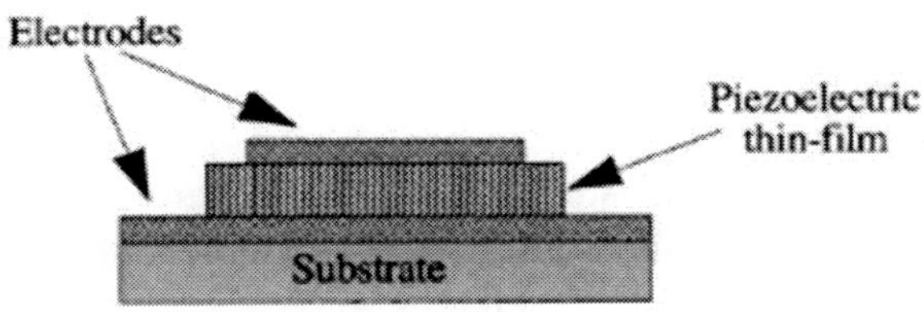

Figure 6.36. Typical configuration of a piezoelectric microactuator.

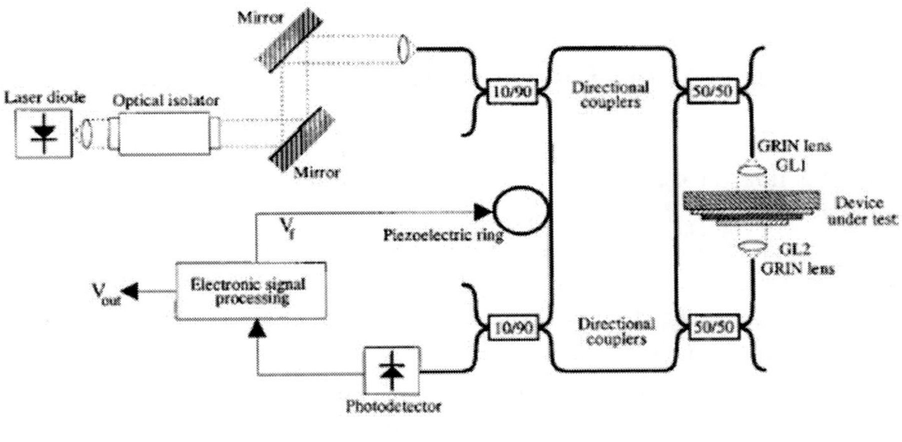

Figure 6.37. Schematic drawing of the interferometer.

In the piezoresponse images shown in Figure 6.34 (b) and (d), domains with opposite polarization exhibit different contrast. Dark regions (negative domains) correspond to domains with polarization oriented towards the substrate, and the polarization in bright regions (positive domains) pointed to the surface of the film. [110] Figure 6.35 shows the local piezoelectric hysteresis loops (piezoloops) of BSPT/Pt and BSPT/LNO measured on a fixed point inside a bright domain (as shown in the cross points in Figure 6.34 (b) and (d)).

From these loops one can observe that under the applied voltage of 10 V, the maximum relative d_{33} value of BSPT/LNO is much higher than that of BSPT/Pt. This indicates that BSPT/LNO exhibit better local and micro piezoelectric properties than BSPT/Pt.

The piezoelectric constant d_{33} is affected by many factors such as the film orientation and residual stress. First, this result is in agreement with what was previously mentioned that the introduction of LNO oxide conductor layer between BSPT film and Pt electrode not only improves the crystalline degree but also remarkably enhances the (100) orientation degree of the ferroelectric films, and further significantly enhances the piezoelectric properties. It is well known that in ferroelectrics, d_{33} shows the maximum value in (100) direction in the MPB phase. [111]

On the other hand, compared to BSPT/Pt, the relatively higher effective d_{33} for BSPT/LNO films may be ascribed to not only the high (100)-orientation degree, but also the modification of the residual stress in BSPT films. [112] The residual stress had been hypothesized to have a significant effect on the piezoelectric film properties. [113] The total residual stress (S_t) consists of three parts:

$$S_t = S_i + S_{th} + S_e,$$

where S_i, S_{th}, S_e are the intrinsic stress, the thermal stress, and the extrinsic stress, respectively. The intrinsic stress is induced by the formation of the grain boundaries as the crystal grain grows and interacts with neighboring grains, etc. Thermal stress is induced by the mismatch between thermal expansion coefficients of the ferroelectric film and the substrate. The extrinsic stress originates from the lattice parameter mismatch between the film and the substrate.

Macro-piezoelectric Properties by Optical Fiber Interferometer Technology

The configuration of a simple piezoelectric microactuator to measure the piezoelectric constant (d_{33}) is illustrated in Figure 6.36. The device consists of two conducting thin films, which form the bottom and top electrodes, and a piezoelectric layer sandwiched between the two electrodes. The actuator makes use of the inverse piezoelectric effect; accordingly, an external electric field is applied between the bottom and top electrodes.

The elastic strain induced in the same direction of the applied field (ε_3) is described by the d_{33} piezoelectric coefficient. The determination of d_{33} requires measurement of the change in thickness induced by the field when the material is under constant mechanical stress. The measurement of such small thickness changes requires high resolution optical techniques, in particular, optical interferometry combined with proper signal detection and processing. The schematic drawing of the optical fiber interferometer is shown in Figure 6.37.

Figure 6.38 shows the measurements of the piezoelectric constant (d_{33}), strain, and phase as the function of applied DC bias voltages of BSPT films on Pt electrodes. These figures indicate that the BSPT films on Pt electrodes exhibit excellent piezoelectric performance. The d_{33} value is as high as 155 pm/V, which is considered to be better than that of PZT films.

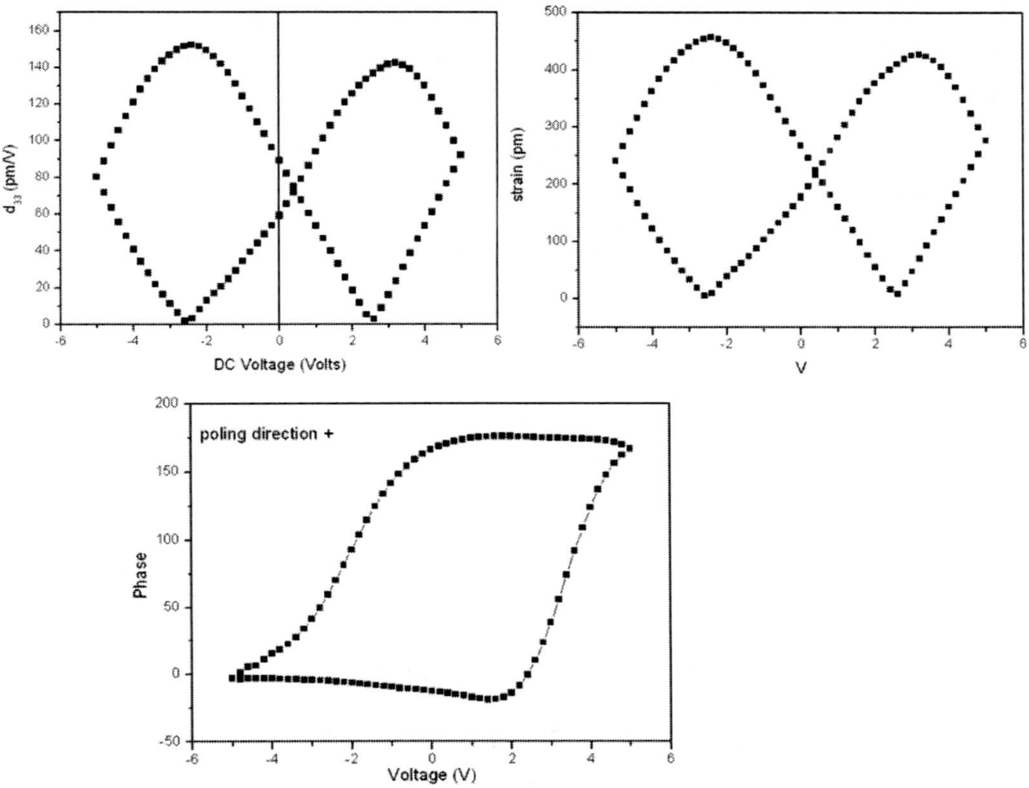

Figure 6.38. the curves of piezoelectric constant (d_{33}), strain, and phase as the function of applied DC bias voltages of BSPT films on Pt electrodes.

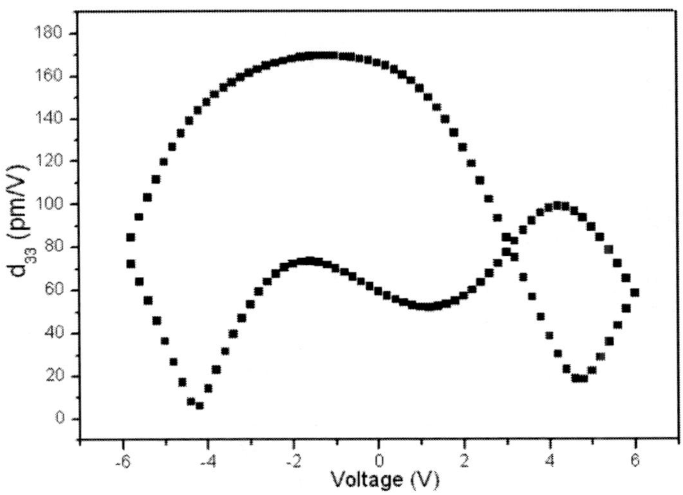

Figure 6.39. Piezoelectric constant of BSPT thin films on $LaNiO_3$ oxide electrodes.

Figure 6.39 shows the piezoelectric constant of BSPT thin films on $LaNiO_3$ oxide electrodes as the function of DC voltages. The obtained d_{33} value is as high as 175 pm/V, which is higher than that of BSPT films on Pt electrodes. This may be due to the fact that the BSPT films deposited on $LaNiO_3$ electrode exhibit higher film orientation degree than the films on Pt electrodes. It is therefore indicated that the increase of the film texture will improve the piezoelectric properties of BSPT thin films.

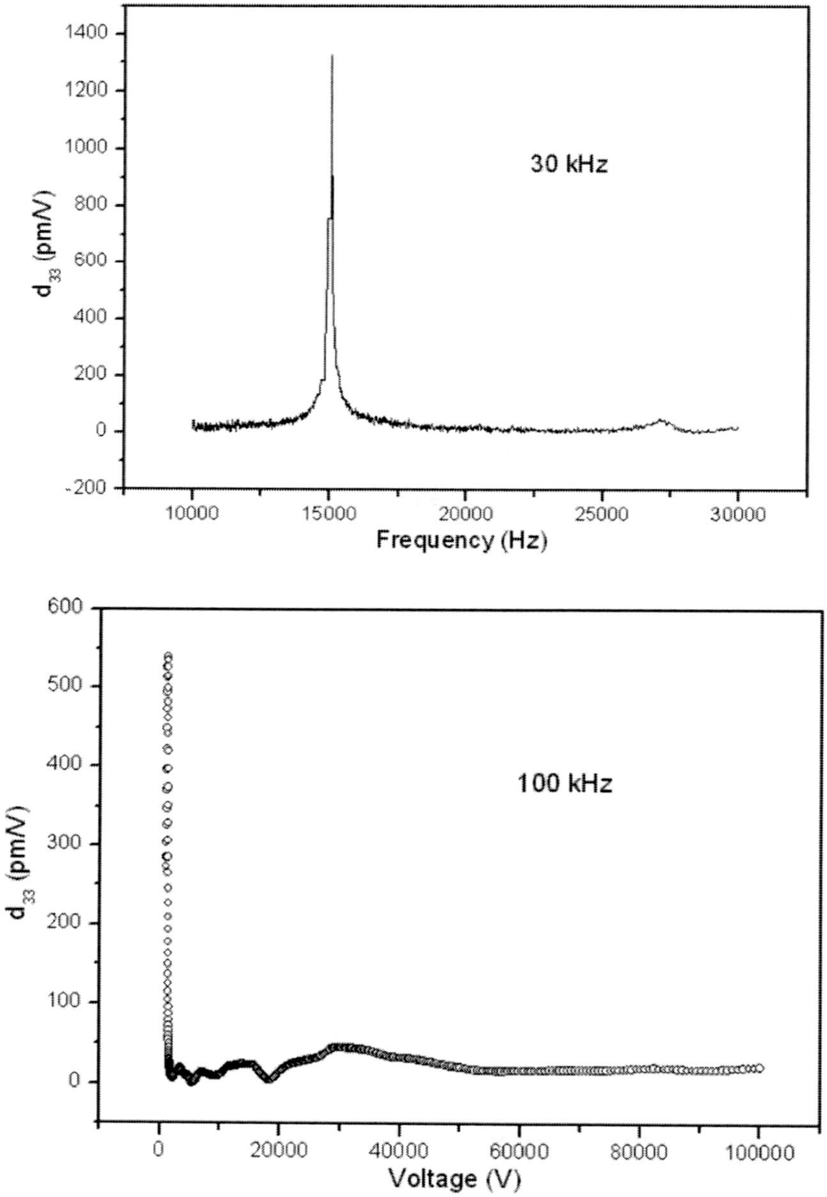

Figure 6.40. frequency dependence of piezoelectric constant of BSPT films on LaNiO3 electrodes.

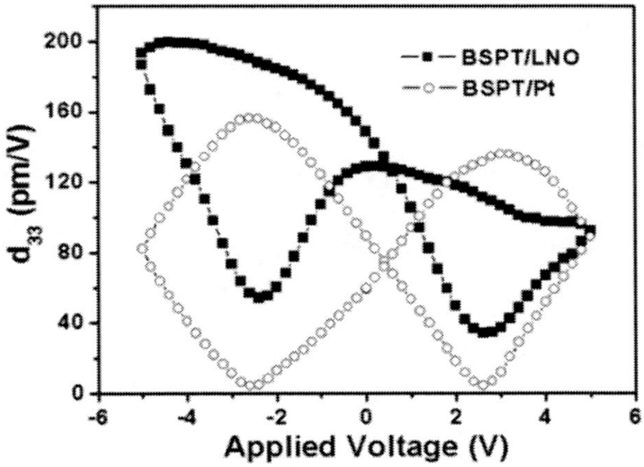

Figure 6.41. The absolute value of the effective piezoelectric coefficient (d_{33}) as a function of applied voltage for BSPT/LNO and BSPT/Pt samples, from which the butterfly-type d_{33} hysteresis loops can be observed.

As shown in Figure 6.40, the piezoelectric constant remains nearly constant at high frequencies (from 40 kHz to 100 kHz). The resonance frequency of the d_{33} value of the films is about 15 kHz.

One great breakthrough obtained in this yearly research is that by the novel optical fiber interferometer measurements and the piezoelectric performance of the BSPT ferroelectric films was proven to be excellent. The piezoelectric constant (d_{33}) of BSPT (37/63) films deposited on the Pt electrode was as high as 155pM/V, which is superior to that of the conventional PZT ferroelectric films (40-120 pm/V). In particular, the high (100) orientation BSPT (37/63) films on $LaNiO_3$ oxide electrode exhibit a d_{33} value of 175 pm/V, which implies the texture engineering and orientation controlling of the BSPT films is beneficial for improving the piezoelectric properties.

The d_{33}-V loops of both samples are asymmetric butterfly shapes (Figure 6.41), and the loop of BSPT/LNO looks more asymmetric than that of BSPT/Pt due to its more asymmetric electrode structure, where the top electrode is metal (Au) and the bottom electrode is an oxide conductor (LNO). In the BSPT/Pt capacitor, both the top and bottom electrodes are metal (Au and Pt), exhibiting a different work function and clamp effect from that of oxide electrodes. In addition, the d_{33} loop of BSPT/LNO shows a sharp shape, which is quite different from that of BSPT/Pt. This may be the result of the contributions of self-polarization clamped by substrate layers and leakage current. The introduction of the LNO electrode can improve the leakage current behavior and decrease the value of the current density, which may also roughly explain why the piezoelectric coefficient loops of BSPT/LNO and BSPT/Pt exhibit different shapes, *ie*., the head of the d_{33} curve is round in BSPT/Pt but sharp in BSPT/LNO.

Most significantly, BSPT/LNO thin films exhibit a much higher piezoelectric coefficient than BSPT/Pt thin films. The maximum of d_{33} in BSPT/LNO thin film achieves about 200 pm/V, while that in BSPT/Pt film it is about 160 pm/V, obtained by means of optical fiber double beam interferometer that can get the macro-piezoelectric coefficients d_{33} with high accuracy. The piezoelectric constant d_{33} is affected by many factors such as the film orientation and residual stress. First, this result is in agreement with what was previously mentioned that the introducing of LNO oxide conductor layer between the BSPT film and the

Pt electrode not only improves the crystalline degree but also remarkably enhances the (100) orientation degree of the ferroelectric films, and further enhances the piezoelectric properties. It is well known that in ferroelectrics, d_{33} shows the maximum value in (100) direction in the MPB phase. The high d_{33} values presented here are also encouraging that MPB BSPT thin film shows much higher piezoelectric response than the highly (100) oriented MPB PZT thin films (d_{33} ~ 80-120 pm/V).

The LNO interlayer inserted between the ferroelectric film and Pt substrate may modify the residual stress in the films. [100-102] (1) In the film fabrication, the LNO interlayer may act as a seed layer and a template of film orientation, leading to improvement in the crystal grain growth and allowing the grains to be well aligned and interacted. This may lead to a greatly decreased intrinsic stress, and may contribute to the decrease of the total residual stress. (2) The thermal expansion coefficients of Pt and LNO are about 27×10^{-6}/K and 8.2×10^{-6}/K, respectively; while that of BSPT can be estimated to be $(1\sim10)\times10^{-6}$/K (for instance, the thermal expansion coefficient of the PZN–PZT was 5×10^{-6}/K). According to this consideration, the LNO can act as a buffer layer of thermal transformation between the BSPT film and the Pt electrode, and decrease the thermal stress in the film of BSPT/LNO. (3) The lattice parameters of Pt (~ 0.3900 nm), LNO (~ 0.384 nm), and BSPT (~ 0.405 nm), make the difference of the lattice mismatch between the film structures of BSPT//LNO and BSPT//Pt very small, from which the contribution of the extrinsic stress to the total residual stress can be neglected. Therefore, the LNO layer introduced between the Pt electrode and the BSPT ferroelectric film can serve as not only a template for optimizing the film orientation, but also a buffer layer for modifying the residual stress, and significantly enhance the electrical performance.

In summary, LNO thin films were successfully prepared on a Pt/TiO_2/SiO_2/Si substrate by a sol-gel technique, for the subsequent sol-gel deposition of BSPT (37/63) thin films. The effects of LNO thin films on the structural and electrical performance of sol-gel derived BSPT thin films were investigated. As an interlayer between the ferroelectric film and Pt bottom electrode, $LaNiO_3$ films served multiple functions of seed layers, oxide electrodes, buffer layers, and textured templates. In comparison to the BSPT thin films deposited on the pure platinized silicon substrates, the ferroelectric thin films prepared on this multifunctional interlayer exhibit highly (100)-orientation, enhancing significantly the dielectric, ferroelectric and piezoelectric properties. This multi-function-layer strategy is anticipated to become a valuable pathway for tailoring the properties in other ferroelectric, magnetic, or superconductive oxide perovskite films integrated on the silicon substrates.

6.5. Piezoelectric Force Microscopy Study of Domain Structure in High Tc Ferroelectric Films

The piezoelectric force microscopy (PFM) technique has become an ideal and unique technique to study ferroelectric materials, through which the local piezoresponse induced ferroelectric domain configurations and their evolution as applied ac electric field can be observed at the nanoscale level. An introduction to principles for studying the ferroelectric domain structure by PFM has been provided, and several results on domain structure and piezoelectric response in high Curie temperature ferroelectric thin films has been introduced.

(1-x)$BiScO_3$-x$PbTiO_3$ (BSPT) thin films are recognized as promising high Curie temperature (Tc) ferroelectrics for applications in MEMs, memory devices, etc., in a high temperature environment. Morphotropic phase boundary (MPB) BSPT exhibits a high Tc 450 °C (100°C higher than PZT) and comparable properties to those of relaxor-PbTiO3 meeting the requirements for high temperature applications in the automotive and aerospace industry, etc. However, there is a lack of understanding on the microscopic mechanism of polarization behavior in high-Tc ferroelectric films. Piezoelectric force microscopy (PFM) has become the ideal tool to directly observe the switching behavior and evaluate local polarization reversal in ferroelectric thin films at the microscopic level, and is an important complement for macroscopic studies. Along with the local piezoelectric measurements and domain visualization, the PFM technique allows for a direct matching of local properties to the microstructures, as both of them can be imaged simultaneously.

This part presents the recent development of high-Tc ferroelectric films, the basic principle of PFM, and some investigation results in local domain structures and polarization behaviors in BSPT thin films.

MPB 0.37$BiScO_3$-0.63$PbTiO_3$ systems (here abbreviated as BSPT (37/63)) were selected for investigation. The sol-gel spin coating technique was applied to deposit BSPT (37/63) thin films. The spin rate ranges from 3000 to 5000 rpm, and the spin time is 30 seconds. BSPT (37/63) films were firstly dried on a hot-stage at the temperature about 350°C for 5 minutes, then pyrolyzed at 650°C for 10 minutes. Subsequently, these films were converted to peroveskite BSPT(37/63) ferroelectric thin films by a conventional thermal annealing (CPA) process, at the crystallization temperature of 750°C for 2 hours. Their domain structure and electrical properties are evaluated through the PFM technique.

Introduction to Piezoelectric Force Microscopy (PFM)

3.1. The Atomic Force Microscope

After the invention of the scanning tunneling microscope (STM) in 1981, [114] the investigation of local properties with atomic-scale resolution became possible. As STM was limited to conductive surfaces, the atomic force microscopy (AFM) was then developed to measure sample topography by employing a tip and force sensor to detect repulsive forces. [115] Many additional techniques have been developed to probe other forces (and hence various film properties) on the nanometer scale (nanoscale), such as piezoresponse force microscopy (PFM) [116] and magnetic force microscopy (MFM). [117]

AFM has several primary components: first, a micro-fabricated cantilever, which serves as a force sensor, a device to detect cantilever deflection, and the second, a scanner which scans either the cantilever or sample stage and adjusts the tip-sample distance. The cantilevers are usually made of Si or other materials, and are coated on the backside with metal to increase reflectivity. At the end of the other side of the cantilever is a sharp probe tip that contacts the sample. A laser and position sensitive photodetector (PSPD) are used to detect cantilever deflection. In AFM, a laser beam is reflected off the backside of a cantilever and directed onto the PSPD, which can detect the changes of cantilever deflection. The third component is either a tip-scanning or sample-scanning mechanism.

The two main working modes for imaging surface topography in AFM are contact and non-contact modes. In the contact mode, the tip is in contact with the sample, and an image is generated by measuring either the tip deflection in a constant height mode or the voltage applied to the scanner in a constant force mode. In the non-contact mode, the tip is kept at a distance of tens to hundreds of Ångstroms from the sample driven (made to oscillate) near its resonance frequency. By detecting the oscillation amplitude of the tip (or feedback required to keep the frequency or tip-sample separation constant), an image of the surface topography can be generated.

PFM Experimental Setup

PFM is a contact mode imaging technique [116, 118]. The setup schematic is shown in Figure 6.42. A modulation or driving voltage is applied between the conductive tip and a back electrode, creating an alternating electric field that causes the piezoelectric response to expand and contract at the modulation frequency, and causes the tip to deflect. The tip deflection at the modulation frequency is measured using a standard lock-in technique. The modulation frequency is kept above the AFM feedback loop response frequency to avoid cross talk between surface topography and the measured piezoresponse, and below the resonant frequency of the cantilever to eliminate resonance between oscillations of the cantilever and the sample. Thus, if the tip deflection is due entirely to the sample deformation as a result of the converse piezoelectric effect, we can measure the piezoresponse of the sample through measuring the deflection.

There are two main approaches to PFM depending on the homogeneity of the applied field. First, when a homogeneous electric field is applied, we can measure the effective longitudinal piezoelectric constant, d_{33}. When the top electrode is larger than the tip-sample contact area, a homogeneous field is generally generated to excite piezoelectric oscillations. In this case, the piezoelectrically excited region is large compared to the tip radius, and when such a top electrode is present, the sample is often referred to as a capacitor, which allows standard polarization, capacitance, and current measurements to be performed. The other approach is to detect the piezoelectric response generated on a bare surface by the conductive tip. This method is advantageous because it allows one to study the material directly; a correlation can be obtained between the film microstructure and the piezoresponse or the domain configuration. This approach also exhibits high resolution features, making it possible to investigate individual domains and to switch (e.g. cause polarization reversal) individual ferroelectric domains and grains. Furthermore, it may be possible to investigate the microscopic mechanism of switching and domain wall motion.

This PFM system can measure both piezoresponse phase and magnitude, as well as measure local piezoelectric hysteresis loops, allowing us to image spatial variations in piezoelectric properties and to measure ferroelectric properties at individual points.

Piezoelectric Measurements in PFM

The measurement of the piezoelectric constant using PFM can be accomplished in a point by point mode, using the values on the digital readout of the lock-in amplifier or by analyzing the PFM magnitude image.

1. Determining the Piezoelectric Constant From the Lock-In [118]

This measurement is performed by lowering the tip (with the AFM scanner) and measuring the tip deflection (with the photodiode) while the tip is in contact with the surface. The scanner must first be calibrated with a known step height in order to calibrate the photodiode signal to a known scanner displacement, and the conversion constant must be measured again anytime the laser position is changed or the cantilever is replaced. The 10kHz component of voltage signal is measured by the lock-in amplifier, which can be converted to a distance using a calibration procedure.

With zero strain and tensor notation dropped for simplicity, the total strain of the film under an applied field, can be considered as:

$$S = \frac{\Delta t}{t} = dE + \gamma E^2,$$

where t is the film thickness and Δt is the change in film thickness.

If we apply a modulation voltage $V = V_0 \sin(\omega t)$, and assume that E=V/t, the sample thickness change is given by

$$\Delta t = \frac{\gamma}{2t} V_0^2 + dV_0 \sin(\omega t) + \frac{\gamma}{2t} V_0^2 \sin(2\omega t - \frac{\pi}{2}).$$

If the tip deflection is due to the sample deformation, the amplitude of the 10kHz component of the change in thickness is

$$\Delta t_{10kHz} = dV_0 = \alpha V_{LI}.$$

Assuming that the modulation voltage is in V RMS, where the photodiode conversion constant is α [Å/mV], and the lock-in voltage output is V_{LI} [mV RMS], the piezoelectric constant (d) is related to the lock-in output voltage by

$$d = \alpha \left(\frac{V_{LI}}{V_0} \right) \qquad [\text{Å/V}],$$

where V_0 is in units of V RMS.

2. Image Ferroelectric Domains by PFM

PFM can be used to monitor the phase of the tip oscillation with respect to the modulation voltage in order to image ferroelectric domains. One of the main advantages of PFM is that it is a nondestructive technique for measuring domain structure. In ferroelectrics, the piezoresponse is directly related to the polarization direction. Through monitoring the phase of the piezoresponse, it is possible to determine the orientation of ferroelectric domains. The sign of the piezoelectric constant and the polarity of non-ferroelectric materials can also be determined from the piezoelectric sign convention.

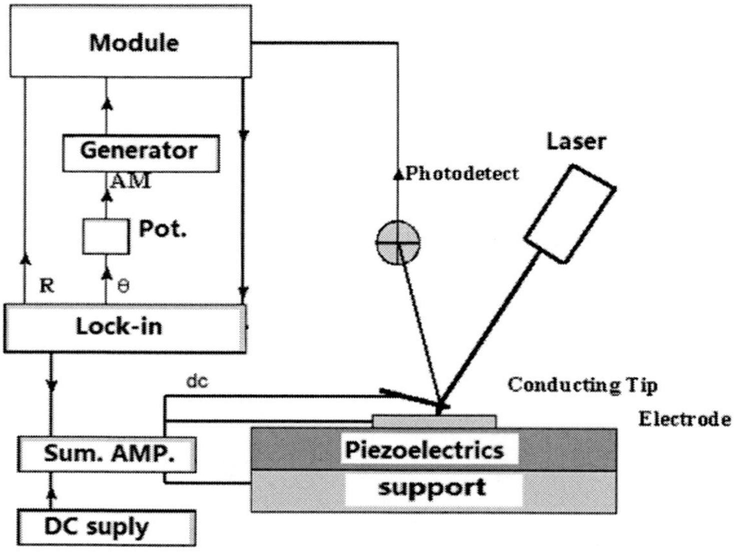

Figure 6.42. Schematic of the PFM technique as applied to piezo/ferroelectrics.

PFM Studies in BSPT High-Tc Ferroelectric Films

1. Investigation of the Evolution of Micro Domain Structure by PFM

The PFM technique for the imaging of polarized regions was applied for the visualization of the domain structure of ferroelectric films, which is based on the detection of local electromechanical vibrations of the ferroelectric sample caused by an external ac field applied between the conducting tip and a counter electrode (Figure 6.42). The deflection signal of the cantilever that oscillates together with the surface of the sample is detected using two lock-in amplifiers and is imaged along with the regular topographic signal in the contact mode.

As described above, the domain structure is visualized by monitoring the first harmonics of the deflection signal. The amplitude is proportional to the local longitudinal piezoelectric coefficient d_{33} and the phase φ reflects the polarization direction (about 0° if the field is parallel to polarization and about 180° if it is antiparallel). If the signal with phase is imaged ($d_{33}\cos\varphi$), the contrast depends on both the amplitude of d_{33} and polarization direction. The bright areas on this image depict polarization directed towards the bottom electrode and dark areas correspond to the domains of opposite direction. The applied voltage has to be high enough in order to cause the measurable signal, but it should be sufficiently small to prevent local switching and domain wall motion.

A modified commercial atomic forcemicroscope (Multimode, Nanoscope IIIA, Digital Instruments) was used in the experiment. A conductive Pt-coated Si tip–cantilever (NSG01/Pt, NT-MDT) system was used for the application of external voltages and for vibration detection. Ferroelectric films were excited by an external ac voltage Vac applied between the PFM tip and the bottom electrode, and the deflection signal from the cantilever was detected by a lock-in amplifier. A topographic image of the film surface was taken simultaneously with the domain image. In the experiments, the films were scanned with the different voltages applied to the tip over the area of 2×2 μm^2.

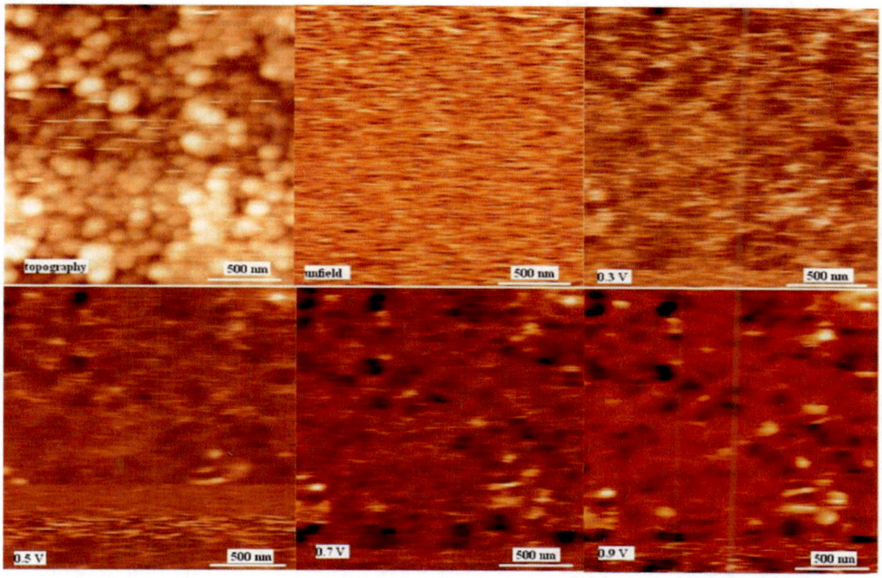

Figure 6.43. (color on-line only) the images of topography and domain structure of unseeded BSPT films on Pt electrode.

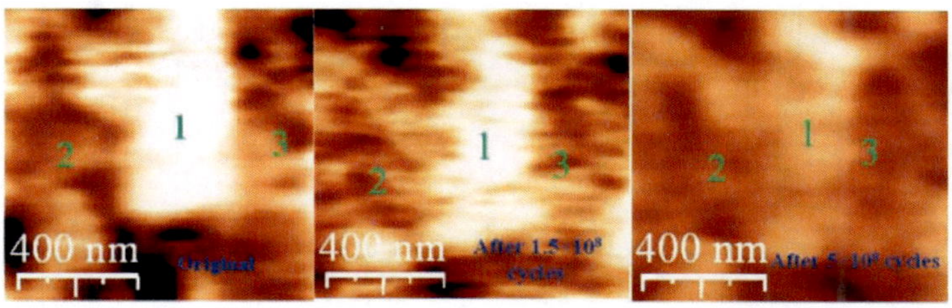

Figure 6.44. (color on-line only) Domain configuration and evolution of unseeded BSPT thin films, by applying different switching cycles: original; 1.5 x 108 switching cycles; 5 x 108 switching cycles.

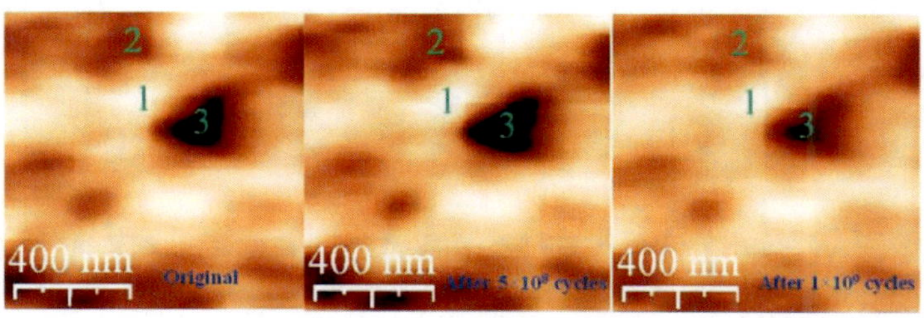

Figure 6.45. (color on-line only) Domain configuration and evolution of PbTiO3 seeded BSPT thin films, by applying different switching cycles: original; 5 x 108 switching cycles; 1 x 109 switching cycles.

2. Micro-scale Ferroelectric Domain in BSPT Thin Films on Pt Electrodes

Figure 6.43 shows the images of topography and domain structure of unseeded BSPT films on Pt electrode taken from PFM. From the two-dimensional morphology image, the

BSPT/Pt film shows a dense grained microstructure. The average grain size is about 150 nm, the surface roughness is about 9 nm.

Piezoelectric domain images acquired simultaneously with the topography of BSPT/Pt films are also shown in Figure 6.37. These images were respectively taken under conditions by applying a small ac voltage (0-0.9 V peak-to-peak, 50 kHz) between the grounded tip and the bottom electrode. The comparison of the topographic and piezoresponse images of the BSPT films allows us to establish an apparent correlation between the surface morphology (grain size) and domain structure. It is observed that the ac field induced ferroelectric domains exhibit a size of 100-150 nm, which is comparable to the grain size.

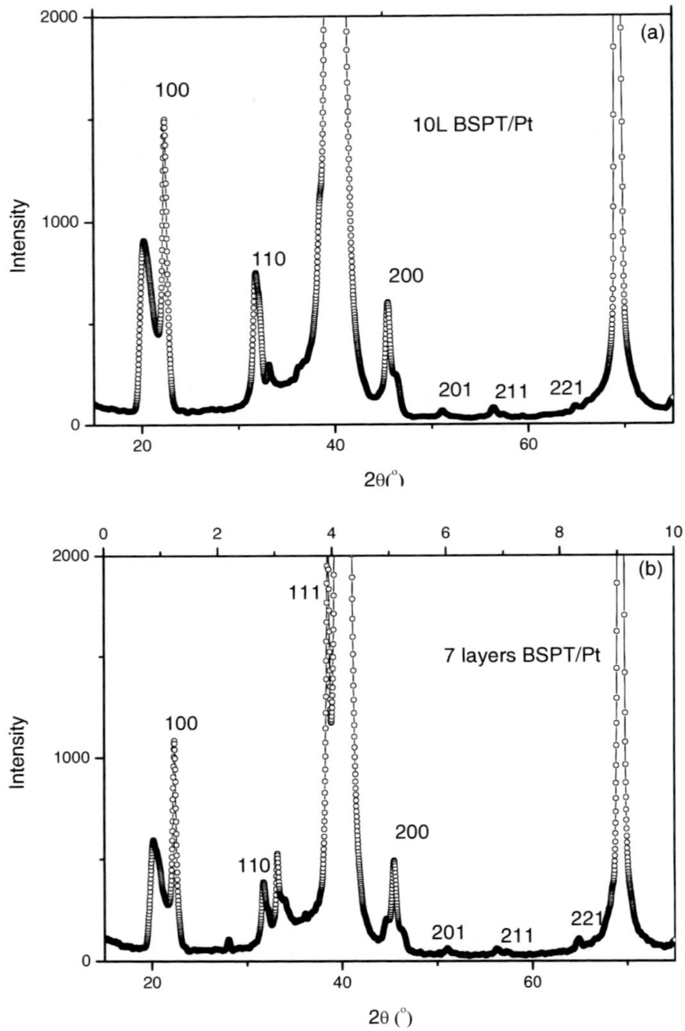

Figure 6.46. XRD patterns of 10 layers (a) and 7 Layers (b) BSPT(37/63) thin films on bare $Pt/TiO_2/SiO_2/Si$ substrate.

In the piezoelectric image, domains with opposite polarities exhibit different contrast. Dark regions (negative domains) correspond to domains with polarization oriented towards the substrate, and bright regions (positive domains) to domains with polarization oriented to the film surface. Since the PFM technique is sensitive only to the component of polarization normal to the film surface, grains with in-plane polarization (with vanishing out-of-plane polarization) will exhibit an intermediate grey contrast. The gray contrast regions can also be attributed to the incomplete 180° switching and/or the formation of 90° domains. With the applied ac voltages increased from 0 V to 0.9 V, the population of the positive and negative domains increase, and the contrast of the ferroelectric domain become clearer.

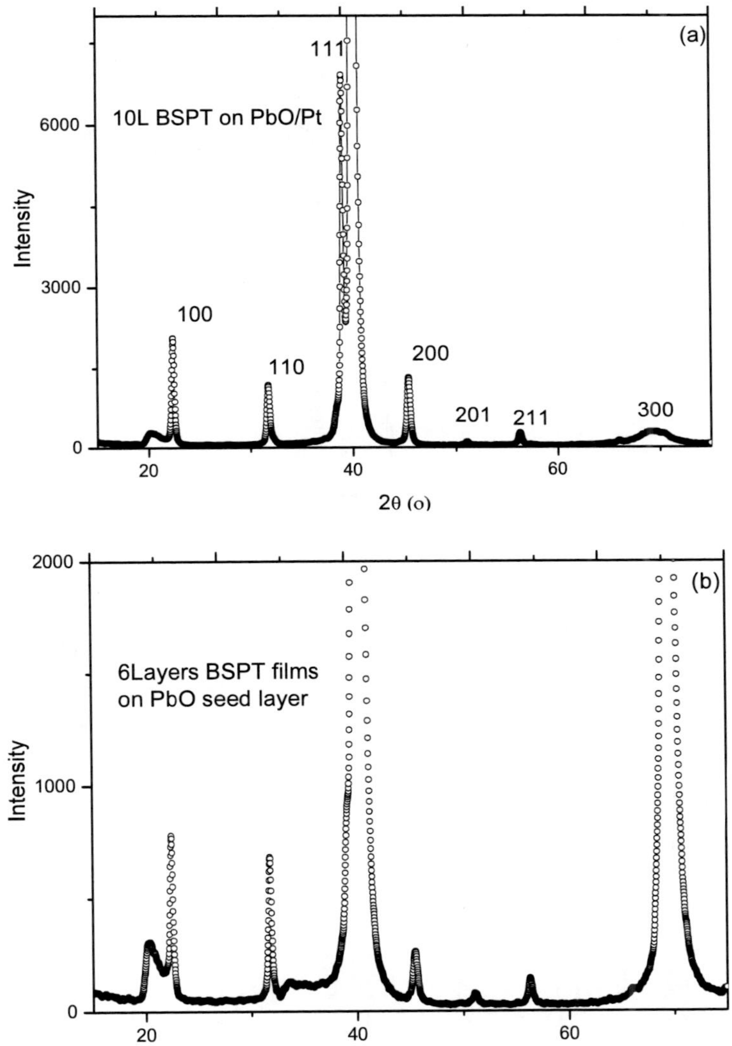

Figure 6.47. XRD patterns of 10 layers (a) and 6 layers (b) BSPT(37/63) thin films on PbO/ Pt/TiO_2/SiO_2/Si substrate.

3. Local Fatigue Characterization by PFM in $PbTiO_3$ Seeded BSPT Films

It is well known that ferroelectric thin films deposited on Pt metal electrodes exhibit serious fatigue of the ferroelectric response, denoted by the reduction of the switchable polarization as a function of the switching cycles, affecting their performance and limiting their utilization. To overcome this problem, one potential effective approach is to introduce a seed layer between the Pt electrode and ferroelectric films. The major objective of this work is to study the effect of the $PbTiO_3$ (PT) seed layer on the micro-scale switching performance and local fatigue behavior of MPB $0.37BiScO_3$-$0.63PbTiO_3$ (BSPT) thin films. The local characterization of piezoelectric force microscopy (PFM) on the domain degradation was conducted, as shown in Figure 6.44 and Figure 6.45.

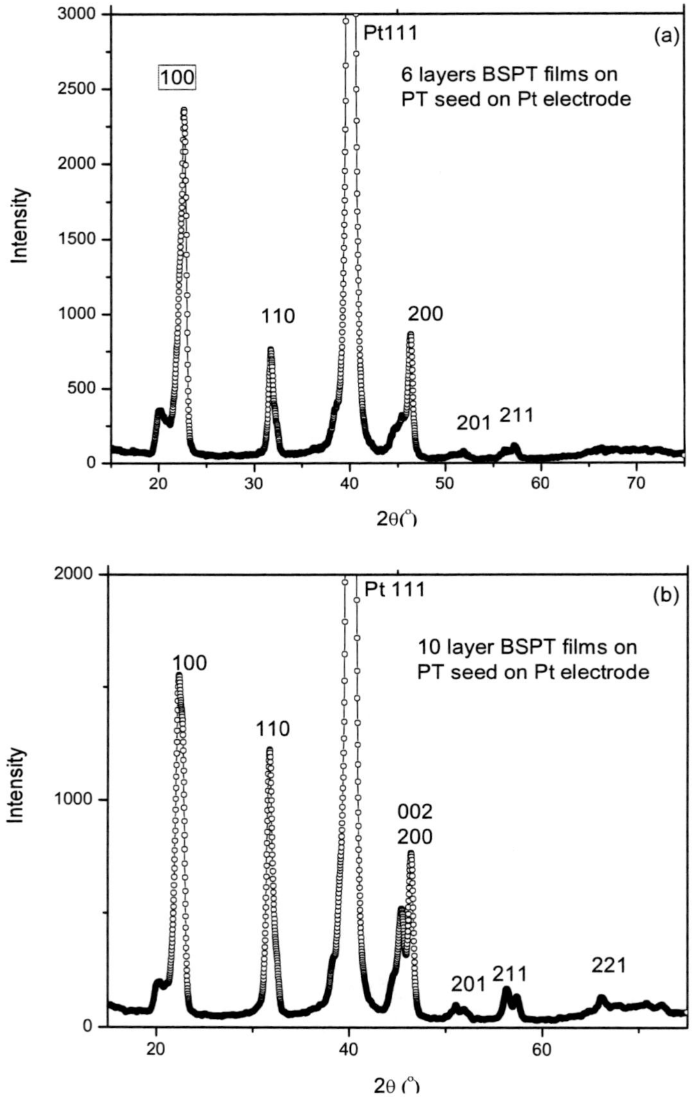

Figure 6.48. XRD patterns of 6 layers (a) and 10 layers (b) BSPT(37/63) thin films on $PbTiO_3$/ $Pt/TiO_2/SiO_2/Si$ substrate.

We found that the PT seed layer improves the film quality and the ferroelectric properties of BSPT thin films deposited on Pt electrodes. At the macroscopic level, the seeded BSPT films show no ferroelectric fatigue until after more than 10^9 switching cycles, while the identical ferroelectric films on bare Pt exhibit obvious fatigue behavior at only 10^7 cycles, which indicates that the introduction of a PT seed layer improves the switching reliability of BSPT thin films on the Pt electrode. At the microscopic level, after switching for 1.5×10^8 cycles, the unseeded BSPT thin films exhibit an evident degradation of the out-of-plane domain images with an increase in the switching cycles. In comparison, the domain configuration in the seeded BSPT thin films do not apparently degrade or disappear. These microscopic level observations clearly imply that the use of PT seed layer improves the polarization switching and fatigue behavior of BSPT films.

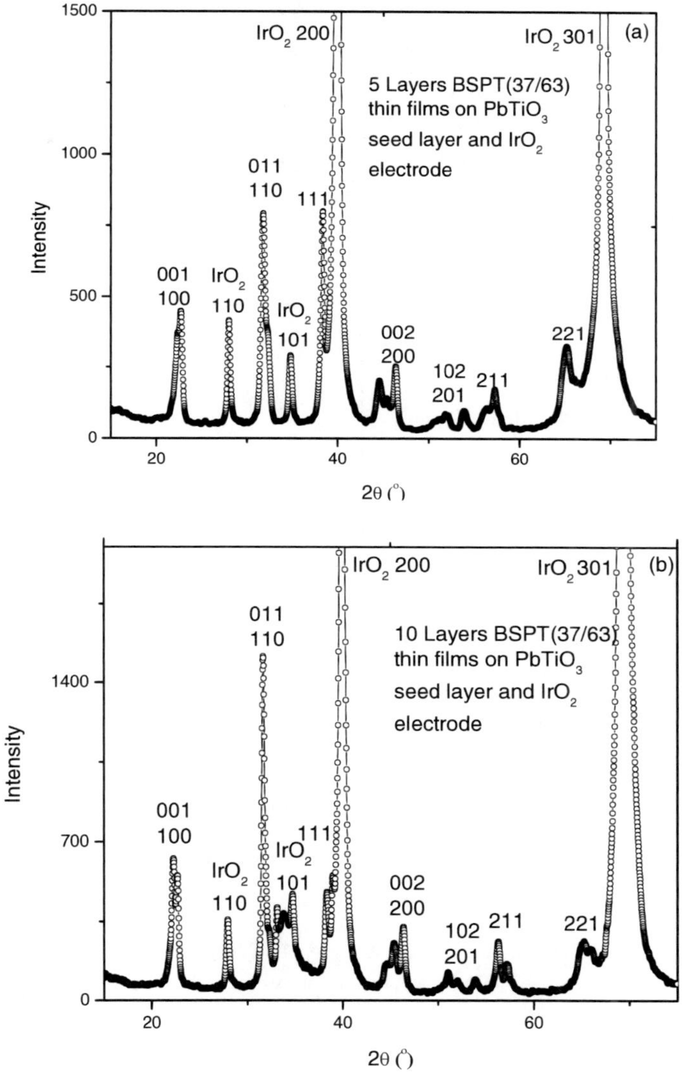

Figure 6.49. XRD patterns of 5 layers (a) and 10 layers (b) BSPT(37/63) thin films on $PbTiO_3$/ IrO_2/TiO_2/SiO_2/Si substrate.

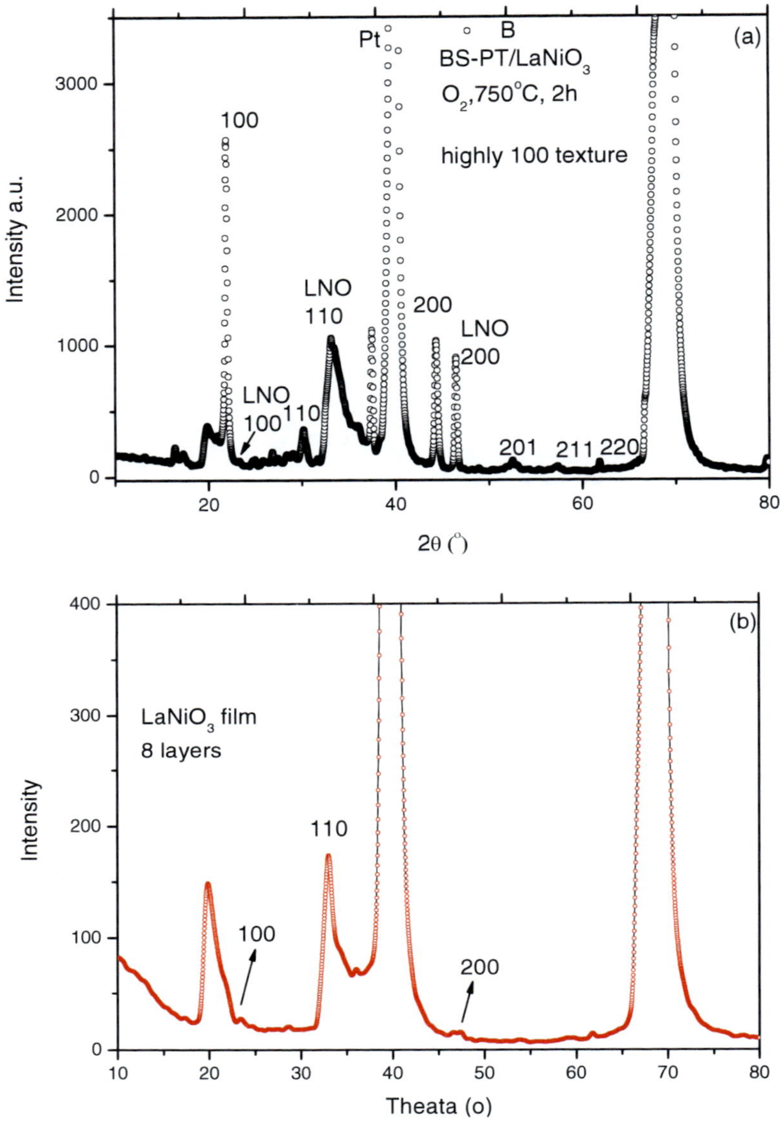

Figure 6.50. XRD patterns of BSPT(37/63) thin films on $PbTiO_3/LaNiO_3/Pt/TiO_2/SiO_2/Si$ substrate (a) and $LaNiO_3$ films on $Pt/TiO_2/SiO_2/Si$ substrate(b).

In summary, an introduction to principles for studying ferroelectric domain structure by piezoelectric force microscopy (PFM) has been provided, and several results on the domain structure and piezoelectric response in BSPT high Curie temperature ferroelectric thin films has been introduced. PFM has become a unique technique to characterize ferroelectric materials. By this technique, the piezoresponse induced ferroelectric domain configurations of BSPT films and their evolution as applied ac electric field was observed at the nanoscale level. These local piezoresponse behaviors will correlate with the macroscopically measured piezoelectric properties of BSPT films.

6.6. TEXTURE CONTROL OF BSPT THIN FILMS

It was reported that, by introducing a seed layer, the crystallization temperature of PZT films could be lowered, and the texture degree of PZT films could be improved [118]. In order to increase the orientation degree and control the texture degree of BSPT thin films, we also introduced a seed layer of $PbTiO_3$ and PbO, respectively. On the other hand, the thickness of ferroelectric film is also a key issue for controlling the texture degree. In this research, the study of film thickness effect on the texture control on BSPT thin film using different seed layer or electrode was conducted.

The XRD curves shown in Figure 6.46 show the primary results of the thickness effect on the XRD patterns of unseeded BSPT thin films. In this case, the thickness of BSPT thin films do not apparently affect the texture control, since all these two thin films deposited on a bare Pt electrode show a strong preference orientation of (100). The increase of thickness only plays a role in enhancing the diffraction intensity of these films.

Figure 6.47 shows the XRD patterns of different layers (thickness) BSPT thin films on the PbO seed layer. In this case, the introduction of the PbO seed layer makes the films show (111) texture, and the intensity of (100) peak is stronger than that of (110) peak. With the thickness increase, the diffraction intensity and crystallinity degree is enhanced, and the (111) texture degree is also increased. In Figure 6.48, it seems that the thickness evidently influences the texture degree of BSPT films on the $PbTiO_3$ seed layer. With the thickness decrease, the thin film shows a stronger (100) preference orientation.

The effect of the oxide electrode on the texture control of BSPT thin films is also investigated. Figure 6.49 and Figure 6.50 show the XRD patterns of BSPT thin films on IrO_2 and $LaNiO_3$ oxide electrodes. It seems the IrO_2 electrode does not influence the texture degree of BSPT thin films. However, the film with an oxide electrode of $LaNiO_3$ exhibits a strong (100) texture degree. Further studies on the relation between electric performance and texture degree of BSPT thin films are planned to be carried out in the future.

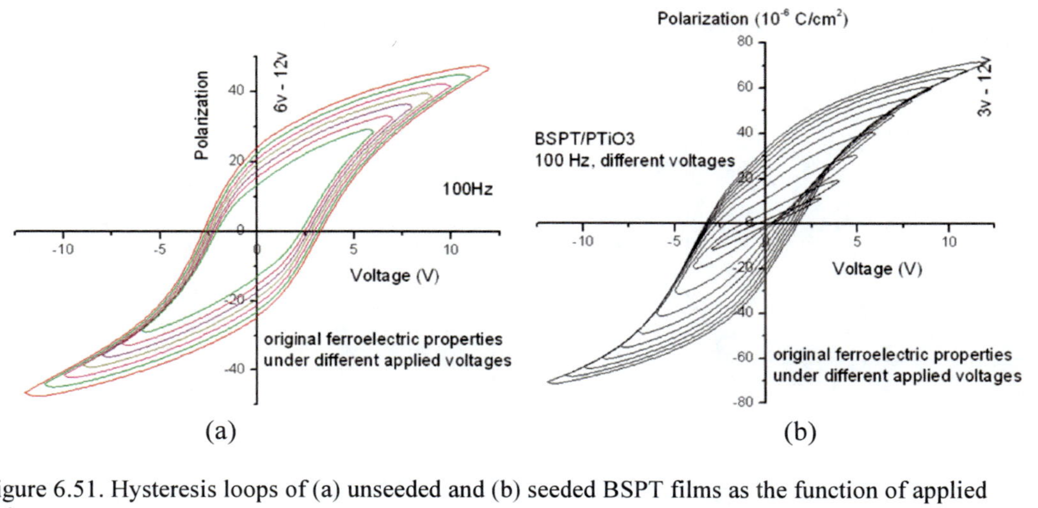

Figure 6.51. Hysteresis loops of (a) unseeded and (b) seeded BSPT films as the function of applied voltage.

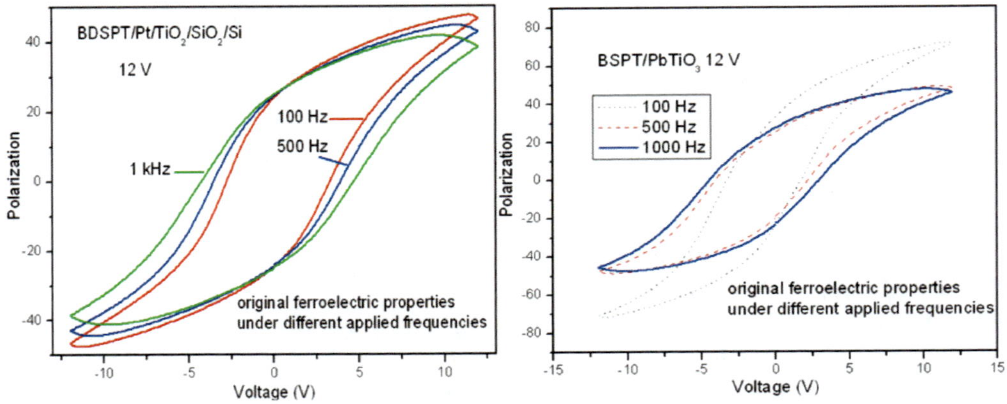

(a) (b)

Figure 6.52. Hysteresis loops of (a) unseeded and (b) seeded BSPT films as the function of frequency.

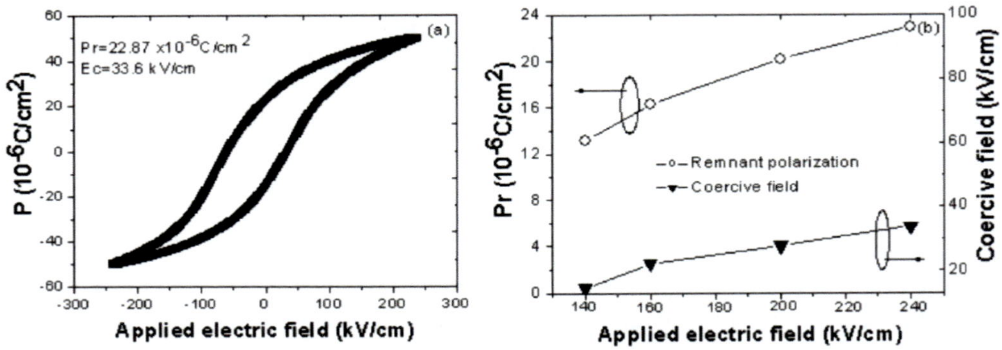

Figure 6.53. (a) Hysteresis loop and (b) variation of remanent polarization and coercive field vs. applied electric field of BSPT 37/63 films with a $PbTiO_3$ seed layer on $Pt/Ti/SiO_2/Si$. The applied voltage is 12V.

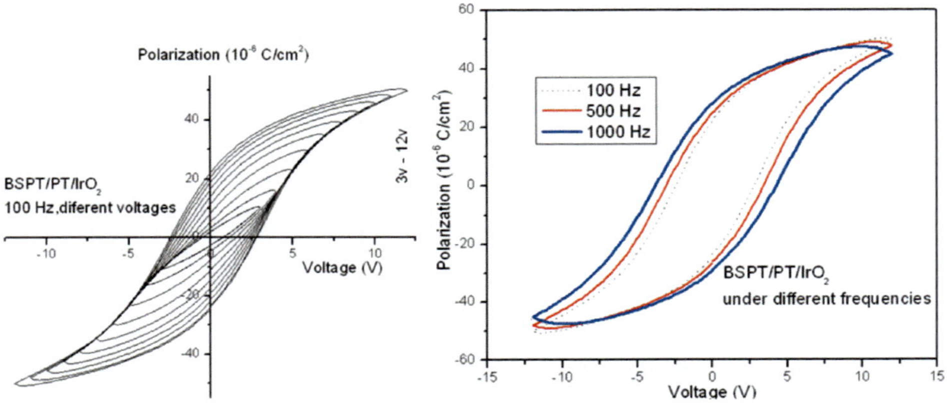

(a) (b)

Figure 6.54. Hysteresis loops of seeded BSPT films as the function of (a) voltage and (b) frequency.

Effect of $PbTiO_3$ Seed Layer on Ferroelectric Properties

Figure 6.51 and Figure 6.52 show the ferroelectric properties of the unseeded and seeded BSPT films as the function of the applied voltage and frequency. Figure 6.53 (a) shows the hysteresis loop of BSPT films with a $PbTiO_3$ seed layer.

Very well defined hysteresis loops were observed with a remnant polarization (P_r) of about 23μC/cm^2 and a coercive field (E_c) of circa 33kV/cm. The variation of remanent polarization and coercive field versus the applied electric field are shown in Figure 6.53 (b), which illustrates a nearly linear variation.

MPB $0.37BiScO_3$-$0.63PbTiO_3$ thin films were fabricated by the sol-gel method with a $PbTiO_3$ seed layer, which exhibit pure perovskite phase and improved crystallinity and better electric properties than unseeded BSPT films. A very low coercive field value of 33 kV/cm shows that the MPB $BiScO_3$-$PbTiO_3$ thin films have potential applications in high temperature ferroelectric memory devices, and will possibly further stimulate the exploration in High-*Tc* ferroelectric films.

Effect of IrO_2 Electrode on the Ferroelectric Properties

Figure 6.54 shows the hysteresis loops of seeded BSPT films on IrO_2 electrodes as the function of voltage and frequency. The ferroelectric hysteresis loops of BSPT/IrO_2 and BSPT/Pt were measured at room temperature (at 12 V) and are shown in Figure 6.55. Very well-defined hysteresis loops are observed for both samples. The hysteresis loop of BSPT/IrO_2 is slightly opened along the polarization axis and exhibits to some extent a higher maximum polarization value, in comparison with that of BSPT/Pt. A remanent polarization (Pr) of 26 μC/cm^2 and a coercive field of about 40 kV/cm were computed for BSPT/IrO_2 films, whereas of about 23 μC/cm^2 and 33.6 kV/cm, respectively for BSPT/Pt ones. IrO_2 electrode slightly increases the remanent polarization as well as the coercive field, but does not substantially affect the ferroelectric behavior of BSPT thin films.

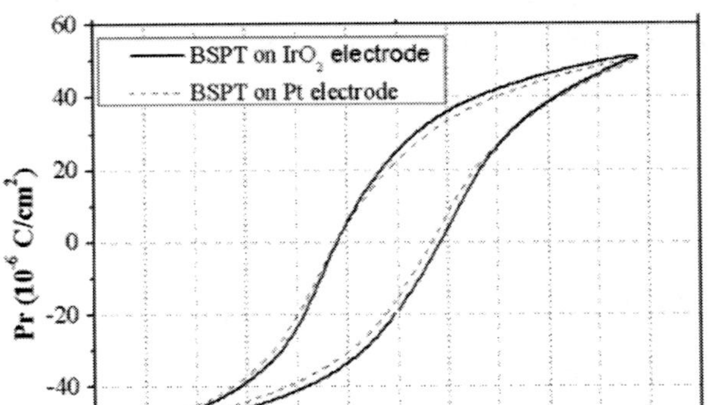

Figure 6.55. Hysteresis loops of BSPT/IrO_2 (solid line) and BSPT/Pt (dashed line) films at 100Hz and at room temperature.

References

[1] Richard E. Eitel, "Novel piezoelectric ceramics:development of high temperature, high performance piezoelectric on the basis of structure", Thesis in Department of Materials Science and Engineering, The Pennsylvania State University, December 2003.

[2] H. Lammer, "Flexible Piezoelectric Films," *International Patent* No. WO 02081401 (October 17 2002).

[3] S. Yoshikawa, A. Bogue and B. Degon, "Commercial Application of Passive and Active piezoelectric Vibration Control, " in Proceedings of the *11th IEEE International Symposium on Applications of Ferroelectrics*, (eds. E. Colla, D. Damjanovic and N. Setter) Montreux, Switzerland, 293-4 (1998).

[4] B. Jaffe, W. R. Cook and H. Jaffe, *Piezoelectric Ceramics,* Academic Press, New York, (1971).

[5] D. A. Berlincourt, C. Molik and H. Jaffe, "Piezoelectric Properties of Polycrystalline Lead Titanate Zirconate Compositions," *Proceedings of the IRE* [2] 220-9 (1960).

[6] S.-E. Park, T.R. Shrout, *J. Appl. Phys*. 82 (1997) 1804.

[7] M. Dong, Z.-G. Ye, *J. Crystal Growth* 209 (2000) 81.

[8] N. Yasuda, H. Ohwa, M. Kume, Y. Hosono, Y. Yamashita, S. Ishino, H. Terauchi, M. Iwata, Y. Ishibashi, *Jpn. J. Appl. Phys*. 40 (2001) 5664.

[9] H.S. Luo, G.S. Xu, Z.W. Yin, *Jpn. J. Appl. Phys*. 39 (2000) 5581.

[10] S. J. Zhang, D. Y. Jeong, Q. M. Zhang, and T. R. Shrout, *J. Cryst. Growth* 247, 131 (2003).

[11] S. J. Zhang, L. Lebrun, S. Rhee, R. Eitel, C. A. Randall, and T. R. Shrout, J. *Cryst. Growth* 236, 210 (2002).

[12] S. J. Zhang, C. A. Randall, and T. R. Shrout, *Jpn. J. Appl. Phys*., Part 2 42, L1152 (2003).

[13] S. J. Zhang, C. A. Randall, and T. R. Shrout, *Appl. Phys. Lett.* 83, 3150 (2003).

[14] V. M. Goldschmidt, *Geochemistry,* Edited by N. F. Mott and S. E. Bullard. Clarendon Press, Oxford, (1954).

[15] R. D. Shannon, "Revised Effective Ionic Radii and Systematic Studies of Interatomic Distances in Halides and Chalcogenides," *Acta. Crystallogr. Sec. A*, 32 [A] 751-67 (1976).

[16] O. Muller and R. Roy : The Maior Ternarv Structural Families, Crystal Chemistry of Non-Metallic Materials, ed. R. Roy, Vol. 4, Springer Verlag, Berlin, Heidelberg, and New York, 221-224, 1974.

[17] N. Yu and V. Venutsev : "Ferroelectrically Active Sublattices and Phase Transitions in Perovskite Type Crystals, " Proc. JSSF-2, Kyoto, *J. Phys. Soc. Jpn*. 49, suppl, B49-52, 1980.

[18] R. E. Eitel, C. A. Randall, T. R. Shrout, P. W. Rehrig, W. Hackenberger, and S.-E. Park, *Jpn. J. Appl. Phys*., Part 1 40, 5999 [2001].

[19] R. E. Eitel, C. A. Randall, T. R. Shrout, and S.-E. Park, *Jpn. J. Appl. Phys*., Part 1 41, 2099 (2002).

[20] Y. Shimojo, R. Wang, T. Sekiya, T. Nakamura, and L. E. Cross, *Ferroelectrics* 284, 121 (2003).

[21] Yoshiyuki Inaguma, et al, "High-pressure synthesis and ferroelectric properties in perovskite-typeBiScO3–PbTiO3 solid solution", *J. Applied Physics*, 95, 231 (2004).

[22] Y.Y. Tomashpol'ski and Y. N. Venettsev, *Translation of Izv. Akad. Nauk SSSR, Neorg. Mat.*, 5, 1297(1969).

[23] R. E. Eitel, et al, Phase Diagram of the Perovskite System (1− x).$BiScO_3$- x$PbTiO_3$, *J. Applied Physics*, 96, 2828 (2004).

[24] V. Porokhonskyy, et al, "Broadband dielectric spectroscopy of (1-x) $BiScO_3$ – x$PbTiO_3$ piezoelectrics", *Aplied Physics Lettters*, 83, 1605 (2003).

[25] Wei Zhao, et al, "Preparation and Characterization of Nanocrystalline (1-x)$BiScO_3$–x$PbTiO_3$ Powder", *J. Am. Ceram. Soc.*, 89 [4] 1200–1204 (2006).

[26] Shujun Zhang, et al, "Characterization of perovskite piezoelectric single crystals of 0.43BiScO3– 0.57PbTiO3 with high Curie temperature", *J. Applied Physics*, 95, 4291 (2004).

[27] Shujun Zhang, "High Curie temperature piezocrystals in the BiScO3-PbTiO3 perovskite system", *Applied Physics Letters*, 83, 3150 (2003).

[28] Shujun Zhang, "Dielectric, piezoelectric and elastic properties of tetragonal BiScO3-PbTiO3 single crystal with single domain", *Solid State Communication*, 131, 41 92004).

[29] Shujun Zhang, "Recent Developments in High Curie Temperature Perovskite Single Crystals", *IEEE Transactions on Ultrasonics, Ferroelectrics, and Frequency Control*, 52, 565(2005).

[30] Takeshi Yoshimura, "Growth and properties of (001) BiScO3–PbTiO3 epitaxial films", *Applied Physics*, 81, 2065 (2002).

[31] E.M. Sabolsky et al, "piezoelectric propersities of <001> textured PMNT ceramics", *Applied Physics Letters*, 17, 2551 (1978).

[32] A. Wu, et al, "Early Stages of Crystallization of Sol-Gel-Derived Lead Zirconate Titanate Thin Films", *Chem. Mater.*, 2003, 15, 1147.

[33] H. Wen, et al, "Characterization of (100)-oriented $BiScO_3$–$PbTiO_3$ thin films synthesized by a modified sol-gel method", *Appl. Phys. Lett.*, 2006, 88 (222904).

[34] J. Scott, C. A. Araujo, *Science* 246, 1400 (1989).

[35] C. A. Araujo, J. Cuchiaro, L. Mcmillan, M. Scott, J. Scott, *Nature*, 374, 627 (1995).

[36] Q. Zhou, K. Shung, Q. Zhang, and F. Djuth, *J. Appl. Phys.*, 97, 104103 (2005).

[37] K. L. Ekinci, M. L. Roukes, *Rev. Sci. Instrum.* 76, 061101 (2005).

[38] K. Kim, S. Lee, *J. Appl. Phys.*, 100, 051604 (2006).

[39] T. R. Shrout, S. J. Zhang, R. Eitel, C. Stringer, C. A. Randall, *14th IEEE International Symposium on Applications of Ferroelectrics*, 23-27, 126(2004).

[40] R. E. Eitel, C. A. Randall, T. R. Shrout, P. W. Rehrig, W. Hackenberger, S.-E.Park, *Japanese Journal of Applied Physics*, 40, 5999 (2001).

[41] Y.Y. Tomashpolski, Y. N. Venetsev, *Translation of Izv. Akad. Nauk SSSR, Neorg. Mat.*, 5, 1297 (1969).

[42] S. J. Zhang, L. Lebrun, S. Rhee, R. Eitel, C. A. Randall, T. R. Shrout, *J. Cryst. Growth* 236 (2002) 210.

[43] S. J. Zhang, C. A. Randall, T. R. Shrout, *Jpn. J. Appl. Phys.*, Part 2. 42 (2003) L1152.

[44] S. J. Zhang, C. A. Randall, T. R. Shrout, *Appl. Phys. Lett.* 83 (2003) 3150.

[45] Richard E. Eitel, *Novel Piezoelectric Ceramics:Development of High Temperature, High Performance Piezoelectric on the Basis of Structure*, PhD Thesis The Pennsylvania State University, December 2003).

[46] R. Eitel, S. Zhang, T. Shrout, C. Randall, *J. Appl. Phys.*, 96, 2828 (2004).

[47] T. Yoshimura, *Appl. Phys. Lett.*, 81 (2002) 2065.

[48] J. Nino, S. Trolier-McKinstry, *Journal of Materials Research*, 19, 568 (2004).

[49] H. Wen, X. Wang, X. De ng, L. Li, *Appl. Phys. Lett.*, 88, 222904 (2006).

[50] H. Wen, X. Wang, X. De ng, t. Sun, L. Li, *J. Appl. Phys.*, 101, 016103(2007).

[51] H. en, X. Wang, L. Li, *J. Am. Ceram. Soc.* 89, 2345 (2006).

[52] H. Wen, X. Wang, L. Li, *J. Appl. Phys.*, 100, 084315 (2006).

[53] H. Wen, X. Wang, C. Zhong, L. Shu, L. Li, *Appl. Phys. Lett.*, 90, 202902(2007).

[54] Y. Li, W. Chen, Q. Xu, J. Zhou, X. Gu, *Materials Letters*, 59, 1361 (2005).

[55] W. Zhu, A. Kholkin, P. Mantas, J. Baptista, *Journal of Materials Science*, 36, 3447 (2001).

[56] T. Oikawa, M. Aratani, H. Funakuba, K. Saito, M. Mizuhira, *J. Appl. Phys.*, 95, 3111 (2004).

[57] Y. Inauguma, A. Miyaguchi, M. Yoshida, T. Katsumata, *J. Appl. Phys.*, 95, 231 (2004).

[58] J. Zeng, M. Zhang, L. Wang, C. Lin, *J. Phys.: Condens. Mater.*, 11, 1139 (1999).

[59] Z. Fu, A. Wu, P. Vilarinho, *Chem. Mater.* 18, 3343 (2006).

[60] J. S. Liu, S. Zhang, H. Zeng, C. T. Yang, Y. Yuan, *Phys. Rev. B.*, 72, 172101(2005).

[61] J. Perez, P.M. Vilarinho, A.L. Kholkin, *Thin Solid Films*, 449, 20 (2004).

[62] Y. Yang, S. Lee, S. Kim, B. Chae, M. Jang, *J. Appl. Phys.*, 1998, 84, 5005.

[63] T. Nakamura, Y. Nakao, A. Kamisawa, H. Takasu, *Appl. Phys. Lett.*, 1994, 65, 1522.

[64] D. Van Genechten, G. Vanhoyland, J. D. Haen, J. Johnson, D. J. Wouters, M. K. Van Bael, H. Van den Rul, J. Mullens, L. C. Van Poucke, *Thin Solid Films*, 2004, 467, 104.

[65] Z. Wang, J. Chu, R. Maeda, H. Kokawa, *Thin Solid Films*, 2002, 416, 66.

[66] H. N. Al-Shareef, B. A. Tuttle, W. L. Warren, T. J. Headley, D. Dimos, J. A. Voigt, D. Nasby, *J. Appl. Phys.*, 1996, 79, 1013.

[67] I. Stolichnov, A. Tagantsev, N. Setter, S. Okhonin, P. Fazan, J. S. Cross, M. Tsukada, *J. Appl. Phys.*, 2000, 87, 1925.

[68] M. Chen, T. Wu, J. Wu, *Appl. Phys. Lett.*, 1996, 68, 1430.

[69] B. S. Kang, D. J. Kim, J. Y. Jo, T. W. Noh, *Appl. Phys. Lett.*, 2004, 84, 3127.

[70] T. Nakamura, Y. Nakao, A. Kamisawa, H. Takasu, *IEEE, Proc. 9th Int.Symp. Application of Ferroelectrics,* 1995, 547.

[71] S. Kim, J. Koo, S. Shin, Y. Park, *Appl. Phys. Lett.*, 2005, 87, 212910.

[72] H. Han, J. Zhong, S. Kotru, P. Padmini, X. Y. Song, R. K. Pandey, *Appl. Phys. Lett.*, 2006, 88, 092902.

[73] C. U. Pinnow, I. Kasko, N. Nagel, T. Mikolajick, C. Dehm, F. Jahnel, M. Seibt, U. Geyer, K. Samwer, *J. Appl. Phys.*, 2002, 91, 1707.

[74] N. P. Barradas, C. Jeynes, R. P. Webb, *Appl. Phys. Lett.*, 1997, 71, 291.

[75] J. C. Nino, S. T. Mckinstry, *J. Mater. Res.*, 2004, 19, 568.

[76] S. Chen, X. Dong, C. Mao, F. Cao, *J. Am. Ceram. Soc.*, 2006, 89, 3270.

[77] B. H. Park, S. J. Hyun, C. R. Moon, Byung-Doo Choe, *J. Appl. Phys.*, 1998, 84, 4428.

[78] M. A. Khan, T. P. Comyn, A. J. Bell, *Appl. Phys. Lett.*, 2008, 92, 072908.

[79] Y. Watanabe, *Phys. Rev. B*, 1998, 57, r5563.
[80] A. I. Dedyk, E. A. Nenasheva, A. D. Kanareykin, J. V. Pavlova, O. V. Sinjukova, S. F. Karmanenko, *J. Electroceramics*, 2006, 17, 433.
[81] S. kanzic, R. Jones, B. Jiang, B. White, P. Chu, D. Taylor, S. Gillespie, *Appl. Phys. Lett.*, 1998, 73, 175.
[82] W. Kanzic, *Phys. Rev. B*, 1955, 98, 549.
[83] G. Choi, H. Tuller, D. Goldshmidt, *Phys. Rev. B*, 1986, 34, 6972.
[84] V. Lo, Z. Chen, *IEEE Transaction on Ultrasonics, ferroelectrics, and Frequency Control*, 2002, 49 , 980.
[85] W. L. Warren, B. A. Tuttle, D. Dimos, *Appl. Phys. Lett.*, 1995, 67, 1426.
[86] Wang, Y.; Ganpule, C.; Liu, B.; Li, H.; Mori, K.; Hill, B.; Wuttig, M.; Ramesh, R. *Appl. Phys. Lett.* 2002, *80*, 97.
[87] Kim, D.; Maria, J.; Kingon, A.; Streiffer, S. *J. Appl.Phys*. 2003, *93*, 5568.
[88] Wu, A.; Vilarinho, P. M.; Reaney, I.; I. M. Miranda Salvado, *Chem. Mater*. 2003, *15*, 1147.
[89] Kwok, C.; Desu, S. *J. Mater. Res.* 1993, *8*, 339.
[90] Suzuki, H.; Koizumi, T.; Kondo, Y., Kaneko, S. *J. Eur. Ceram. Soc.* 1999, *19*, 1397.
[91] Zeng, J.; Zhang, M.; Wang. L.; Lin, C. *J. Phys.: Condens. Matter.* 1999, *11*, 1139.
[92] Muralt, P. *J. Micromech. Microeng*. 2000, *10*, 136.
[93] Yang, Y.; Lee, S.; Kim, S.; Chae, B.; Jang, M. *J. Appl. Phys.* 1998, *84*, 5005.
[94] Nakamura, T.; Nakao, Y.; Kamisawa, A.;Takasu, H. *Appl. Phys. Lett.* 1994, *65*, 1522.
[95] Genechten, D.; Vanhoyland, G.; Haen, J.; Johnson, J.; Wouters, D. *Thin Solid Films* 2004, *467*, 104.
[96] Wang, Z.; J. Chu, R. Maeda, H. Kokawa, *Thin Solid Films* 2002, *416*, 66.
[97] Taylor, D.; Damjanovic, D. *Appl. Phys. Lett.* 2000, *76*, 1615.
[98] Ramesh, R.; Schlom, D. *Science*. 2002, *296*, 1975.
[99] Ranjith, R.; Chaudhuri, A.; Krupanidhi, S.; Victor, P. *J. Appl. Phys*. 2007, *101*, 104111.
[100] Chen, M.; Wu, T.; Wu, J. *Appl. Phys. Lett.* 1996, *68*, 1430.
[101] Kang, B.; Kim, D.; Jo, Y.; Noh, T. *Appl. Phys. Lett.* 2004, *84*, 3127.
[102] Nakamura, T.; Nakao, Y.; Kamisawa, A.; Takasu, H. *IEEE, Proc. 9th Int. Symp. Application of Ferroelectrics,* 1995, *9*, 547.
[103] Cho, C., Francis, L.; Polla, D. *Mater. Lett.* 1999, *38*, 125.
[104] Bao, D.; Wakiya, N.; Shinozaki, K.; Mizutani, N.; Yao, X. *Appl. Phys. Lett.* 2001, *78*, 3286.
[105] Ahluwalia, R.; Lookman, T.; Saxena, A. *Phys. Rev*. B 2005, *72*, 014112.
[106] Lotgering, F. *J. Inorg. Nucl. Chem*. 1959, *9*, 113.
[107] Li, Y., Hu, Z.; Sun, J.; Meng, J.; Chu, *J. Appl. Phys. Lett*. 2008, *92*, 042901.
[108] Ramesh, r.; Inam, A.; Wilkens, B.; Chan, W.; Sands, T. *Appl. Phys. Lett.* 1991, *59*, 1782.
[109] Lin, A.; Hong, X.; Wood, V.; Verevkin, A.; Ahn, c. *Appl. Phys. Lett*. 2001, *78*, 2034.
[110] Gruverman, A.; Auciello, O; Tokumoto, H. *Annu. Rev. Mater. Sci*. 1998, *28*, 101.
[111] Du, X.; Zheng, J.; Belegundu, U.; Uchino, U. *Appl. Phys. Lett.* 1998, *79*, 2421.
[112] Lin, A.; Hong, X.; Wood, W.; Verevkin, A.; Ahn, C. *Appl. Phys. Lett*. 2001, *78*, 2034.
[113] Lian, L.; Sottos, N. *J. Appl. Phys*. 2000, *87*, 3951.
[114] Binnig. G., Rohrer. H., Gerber. C., Weibel. E., *Phys. Rev. Lett*. 1982; 49: 57.

[115] Binnig. G., Rohrer. H., Gerber. C., *Phys. Rev. Lett.* 1986; 56: 930.
[116] Güthner. P., Dransfield. K., *Appl. Phys. Lett.* 1992; 61: 1137.
[117] Philip C. D. H., David W. A., Wickramasinghe. H. K., *Appl. Phys. Lett.* 1989; 55: 2357.
[118] Rodriguez. B. J., PhD thesis, Nanoscale Investigation Of The Iezoelectric Properties Of Perovskite Ferroelectrics and III-Nitrides, North Carolina State University, 2003; 81-84.

INDEX

D

E

F

G

N

O

P

Q

R

S

T

U

V

W

X

Y